Reciprocity in
Population Biobanks

Reciprocity in Population Biobanks

Relational Autonomy and the Duty to Inform in the Genomic Era

Ma'n H. Zawati

McGill University, Faculty of Medicine and
Health Sciences, Department of Human
Genetics, Centre of Genomics and Policy
Montreal, Canada

ACADEMIC PRESS
An imprint of Elsevier

Academic Press is an imprint of Elsevier
125 London Wall, London EC2Y 5AS, United Kingdom
525 B Street, Suite 1650, San Diego, CA 92101, United States
50 Hampshire Street, 5th Floor, Cambridge, MA 02139, United States
The Boulevard, Langford Lane, Kidlington, Oxford OX5 1GB, United Kingdom

Notices
Knowledge and best practice in this field are constantly changing. As new research and
experience broaden our understanding, changes in research methods, professional
practices, or medical treatment may become necessary.

Practitioners and researchers must always rely on their own experience and knowledge in
evaluating and using any information, methods, compounds, or experiments described
herein. In using such information or methods they should be mindful of their own safety
and the safety of others, including parties for whom they have a professional responsibility.

To the fullest extent of the law, neither the Publisher nor the authors, contributors, or
editors, assume any liability for any injury and/or damage to persons or property as a
matter of products liability, negligence or otherwise, or from any use or operation of any
methods, products, instructions, or ideas contained in the material herein.

Library of Congress Cataloging-in-Publication Data
A catalog record for this book is available from the Library of Congress

British Library Cataloguing-in-Publication Data
A catalogue record for this book is available from the British Library

ISBN: 978-0-323-91286-0

For information on all Academic Press publications visit our website at
https://www.elsevier.com/books-and-journals

Publisher: Andre Gerhard Wolff
Acquisitions Editor: Peter B. Linsley
Editorial Project Manager: Tracy I. Tufaga
Production Project Manager: Niranjan Bhaskaran
Cover Designer: Alan Studholme

Typeset by TNQ Technologies

To my father, Hilmi M. Zawati, and my mother, Ibtisam Mahmoud, for their lifelong inspiration.

To my wife, Mays Abu Dabasi, and my son, Amir, for their unwavering support and unconditional love.

Contents

Preface

Criticism of the individualistic conception of autonomy is not new. Over a number of years, a great deal of ink has been spilled grappling with its conceptual limitations as well as with solutions aimed at palliating them in the clinical setting. However, much less has been written on the shortcomings of individualistic autonomy in the research field (and even less in the context of population biobanks), where the standard of disclosure of researchers is, according to Canadian courts, more exacting than the standard imposed on clinicians. Similarly, reciprocity is not, in itself, a novel concept, and has been presented in several economics, sociological, and medical analyses. Against this backdrop, this book's original scholarship lies in its use of reciprocity as both a framework to abate limitations of individualistic autonomy in the research setting as well as a conceptual basis for accurately describing, acknowledging, and sustaining the multiple relations at the core of a more relational conception of autonomy in the context of population biobanking. Moreover, by asserting that reciprocity is an appropriate grounding for relational autonomy, this book also demonstrates that reciprocity is a more plausible conceptual basis from which to ground the standard of disclosure in population biobanks.

Acknowledgments

Undertaking doctoral studies can be a lonely business. While one can sometimes feel a degree of isolation during the doctoral cycle, reaching the end of the journey would not be possible without a great deal of support along the way. I am certainly not in a deficit of gratitude, but rather in a surplus of debt to all those without whom this book, which is based on my doctoral thesis at McGill University's Faculty of Law, would not have materialized.

I am particularly indebted to Prof. Lara Khoury, my doctoral supervisor. Without her advice, meticulousness, and constructive criticism, this thesis would not have been possible. In his well-known essay, *"Politics and the English Language,"* George Orwell exhorted his readers to use English words only as keywords and never as shorthand. He advocated for using language in a way that recognizes its power, avoiding short, clever modes of expression. With great care, Prof. Khoury made sure I did exactly that along the way. I thank her for her availability and, especially, for her patience. In the same vein, I would like to acknowledge the support I have received from Prof. Bartha Maria Knoppers and Prof. Alana Klein, both members of my Doctoral Advisory Committee. Their comments and suggestions on my earlier drafts were invaluable. Prof. Knoppers, Director of the Centre of Genomics and Policy (CGP), has been a mentor for me for more than a decade now and I owe her a lot.

I have been privileged to receive comments from Prof. Yann Joly, Research Director at the CGP, on earlier drafts of some chapters. His generous guidance on both a professional and personal level has been priceless. Prof. Michael Parker, Director of the Ethox Centre at Oxford University, was kind enough to host me for 3 months at his centre, where I was fortunate enough to receive comments from him as well.

I would like to acknowledge the generous funding provided to me by the Fonds de recherche du Québec — Société et culture (FRQSC) through a doctoral scholarship. I was also humbled to receive the following awards: Queen Elizabeth II Diamond Jubilee Scholarship, Fellowship from the Research Group on Health and Law, John and Edmund Day Award, Justice K. Hugessen Fellowship, Rathlyn Foundation Award, and a Graduate Excellence Fellowship from the Faculty of Law.

I am particularly indebted to Michael Lang, Academic Associate at the Centre of Genomics and Policy, for his extremely able assistance in updating the chapters of this book. His availability and dedication were invaluable. I would also like to thank Ana Eliza Bonilha for her meticulous help with the references.

This book would not have been written without the encouragement and loving care of my wife, Mays. I thank her for her friendship, patience, and unending support. My father, my mother, and my two brothers have accompanied me for a good portion of this journey. They had to get by with an absent son and an absent brother as the thesis, and ultimately this book, was in progress. I thank them for their under-standing. To my father and mother, I thank them for their constant blessings. In the year I was updating my thesis for eventual publication as a book, I was blessed with the birth of my son, Amir. May he, like his father and grandfather before him, be illuminated by unflinching curiosity on the strenuous, but ever-rewarding pathway to knowledge.

General introduction[1]

> *"Data! Data! Data!" he cried impatiently.*
> *"I can't make bricks without clay."*
> **Sherlock Holmes (in Arthur Conan Doyle,**
> ***The Adventure of the Copper Beeches*)**

The principle of autonomy has been the cornerstone of the physician's duty to inform since paternalistic medical practices receded in the second half of the 20th century [1–7]. Prior to this, physicians would often withhold relevant information from patients in an ostensible effort to protect them from harm [1,8,9]. Up until 1981, for example, the Quebec *Code of Ethics of Physicians* still permitted the medical therapeutic privilege, allowing physicians to "conceal a fatal or grave prognosis from a patient" in the presence of a "valid reason" [9]. Later in the century, health care professions began considering whether withholding information could result in greater harm, on balance, than disclosure [10]. Illustrating this trend, one author notes that in less than 2.decades studies of physician attitudes have shifted from a trend of withholding cancer diagnostics—90% of physicians in 1961—to a general preference for disclosing them—97% [7]. This consideration features centrally in the principle of autonomy. In the medical field, autonomy may be characterized as the right of a patient to make an informed decision without the unjustified interference of others [11]. On one prominent interpretation, respect for autonomy in this context entails giving

> *weight to autonomous persons' considered opinions and choices while refraining from obstructing their actions unless they are clearly detrimental to others. To show lack of respect for an autonomous agent is to repudiate that person's considered judgments, to deny an individual the freedom to act on those considered judgments, or to withhold information necessary to make a considered judgment, when there are no compelling reasons to do so [12].*

In particular, the disclosure of information has become a critical element of the principle of autonomy. One prevalent conception in the medical field claims that the extent of the duty to inform (and, by extension, the duty to disclose) is inversely proportional to an intervention's expected therapeutic benefit. For example, the duty to disclose is typically heightened in cosmetic surgery, organ donation, and research, where individuals are not expected to benefit therapeutically [13,14]. More precisely, Canadian courts have maintained that research participants are entitled to a "full and frank disclosure" [15] during the consent process and that the duties of

---

[1]Portions of this Introduction have previously appeared in Ma'n H Zawati, "There Will be Sharing: Population Biobanks, the Duty to Inform, and the Limitations of the Individualistic Conception of Autonomy" (2014) 21 Health LJ 97.

researchers in this respect are more demanding than the duties physicians owe their patients in a clinical setting [16]. Since research is generally seen as "an undertaking intended to extend knowledge through a disciplined inquiry and/or systematic investigation" [17], courts have reasoned that research participants are not in a therapeutic relationship and, as a consequence, do not stand to benefit in the way that patients in a clinical setting would benefit. This distinction, according to judicial interpretation, necessitates a more exacting duty to inform, one in which researchers are required to provide participants a full and frank disclosure "of all risks, no matter how remote, as well as all other material information about the research" [16] during the consent process.

As research becomes increasingly longitudinal (analyzed and accessed over time), international (crossing boundaries and legal jurisdictions) [18,19], and less directly focused on individuals, the feasibility of applying this standard is being challenged. In addition, research has come to rely less on direct interventions and ever more on cutting-edge bioinformatics technologies capable of generating, curating, and interpreting massive amounts of data [20]. This is especially true in the case of population biobanks, which aim to study data and samples collected on the scale of entire populations over an extended period [21]. Because the law contends that such projects do not have therapeutic aims, they attract a more exacting standard of disclosure during consent. But owing to the nature of population biobanks, there are limitations on the kind of information that may practically be disclosed to research participants. For example, the only information that can be provided to participants on the nature of population biobank research is that the goal is the establishment of the biobank as a resource for future research in health and genomics with ethics approval for subsequent specific projects [22]. Providing a full disclosure to participants enrolled in population biobanks could be difficult given that their data and samples will be used for future yet unspecified research projects.

Over the last decade, much has been written on the kind of consent required in population biobank projects. A number of authors have considered whether broad consent—a model in which participants are informed that their data and samples will be used for future, as-yet unspecified research [23]—satisfies the legal and ethical requirements of informed consent [24,25]. This approach, in opposition to more specific consent, is adopted when the possible uses of data and samples are not identified at the beginning of the relevant project. Broad consent is generally paired with ongoing communication between biobank researchers and participants, in addition to internal (e.g., in-house access committees) and external (e.g., research ethics boards) oversight mechanisms aimed at protecting the rights of participants. Recent regulatory advances, most notably the entering into force of Europe's *General Data Protection Regulation* (GDPR), have placed a renewed spotlight on the conditions under which broad consent would be legally and ethically valid [26]. The GDPR relaxes prior requirements for specific consent, "allowing use of broad consent whenever required by the intended research purposes" [27]. While these discussions on the nature of the consent applicable in biobanking are important, the majority of authors focus on *operational* concerns, examining the governance and practicability of specific and broad consent approaches in the population biobank context, rather than considering the theoretical underpinnings that support the kinds of consent under consideration [28,29]. Perhaps this is why—despite a large number of articles having been written on the topic of consent in biobanking—some continue to argue that consent issues in the field remain unresolved [30].

Over the course of this book, I will refer to the existing literature on the governance and practicability of consent approaches in population biobanking. This literature, however, will not be my central focus. Instead, I will primarily concentrate on what I conceive to be the foundational problem in population biobank consent: the exacting character of the researcher's duty to inform. The rationale supporting this exacting standard, I argue, is both conceptually problematic and practically at odds with the reality of observational research. More precisely, the duty to inform—as it has traditionally been conceived by Canadian courts—focuses on the interests of individual participants while neglecting to consider the interests and significant roles played by the myriad of other stakeholders implicated in the population biobank research. Under the prevailing judicial interpretation, participants are conceived as fully independent agents rather than interdependent with other stakeholders. This approach motivates an exacting standard, one that is not only difficult to meet in the longitudinal observational research context but may also negatively affect the outcome of a research study.

More specifically, when considering the conception of autonomy that is most appropriate when consenting research participants enrolled in population biobanks, I will argue that reciprocity-based relational autonomy adequately plays this role. It does so largely because it is capable of accounting for the numerous complex, ongoing, and multilateral relationships established by population biobank projects. To do so, I will first demonstrate that the current jurisprudential interpretation of the duty to inform in Canada has individualistic autonomy at its core (also referred to as "individual autonomy" in this book). Secondly, I will outline the multiple limitations of individualistic autonomy in the context of population biobanks. These limitations are twofold: first, individualistic autonomy fails to recognize the complexities of benefit considerations in the research setting. Second, given its unidirectional aims (that is to say, an interaction between the participant and another will focus on the participant), individualistic autonomy fails to acknowledge the multilateral relationships necessarily implicated in population biobanking research, including those that implicate the broader research community and the general public. I will then demonstrate how most solutions proposed in the literature to palliate individual autonomy's shortcomings do not resolve the limitations identified above. In doing so, I will pay special attention to the alternative approaches of deliberative autonomy, principled autonomy, the duty to participate in research, and relational autonomy. I will argue that the latter represents the most suitable conception of autonomy in population biobanks. Using theoretical discussions, I will argue, however, that relational autonomy will need to be situated in a conceptual framework that practically describes, acknowledges, and sustains the multilateral relationships found in this species of research, without also compromising the rights of participants. I will demonstrate that the concept of reciprocity can provide such a conceptual basis for conceiving of the multiple relations at the core of relational autonomy in the context of population biobanking. Indeed, I will argue that in spite of certain limitations, reciprocity—a concept motivated by the view that individuals will "help or benefit others at least in part because [they] have received, will receive, or stand to receive beneficial assistance from them" [31]—is an appropriate grounding for relational autonomy and a better conceptual basis from which to frame the disclosure of information during the population biobank consent process.

In order to demonstrate these points, I will mainly focus on the correlative conception of autonomy that the traditional duty to inform exteriorizes and ultimately aims to respect. While this will not prevent me from referring to the duty to inform of researchers from time to time, mainly approaching my analysis at the level of autonomy (rather than consent or the duty to inform) will allow me to study the relations that are at the heart of the conception of autonomy as it has been understood by the courts. This, in turn, will allow me to critically assess whether these relations can also apply to population biobanks, an issue which is at the heart of my book. Using an analogy from the field of genetics, I am interested in the "genotype" in order to understand the "phenotype." While the phenotype is a set of observable characteristics [32] (in this case, how the duty to inform is interpreted by the courts), the genotype (the conception of autonomy and associated relationships) is the underlying part and the focus of my analysis [32]. Approaching the discussion in this way permits me to begin the work of developing a precise alternative conceptual model for autonomy without being limited to a superficial discussion focused solely on a need to provide practical solutions when considering the disclosure of information to participants. Following an examination of the proposed conceptual model for autonomy, I will very briefly turn to the ways in which this new conception may be exteriorized by researchers when disclosing information to research participants (see Fig. 1).

Throughout this book, I have chosen to focus on population biobanks. There are two reasons for this decision. First, population biobanks reflect the complexity of modern research typology. By "typology" I mean to refer to the variety of research projects that presently exist. Giving particular attention to research typology means both that the context in which research is conducted must be considered and that the fact that research is not homogenous will be respected. Indeed, clinical trials differ markedly from population biobanks. Even among biobanks themselves, disease-specific biobanks differ in relevant ways from population biobanks. Each type of project encapsulates different goals, varying methods of recruitment, different researcher—participant relationships, and dissimilar levels of access to data and samples [33,34]. Relying on generalizations (that is, referring to biobanks in general, rather than to specific types of biobanks) in discussions of particular issues runs the risk of failing to capture all of the intrinsic characteristics of the biobank under study and how best the unique issues it presents can be contemplated. For this reason, a singular focus on population biobanks permits me to avoid such

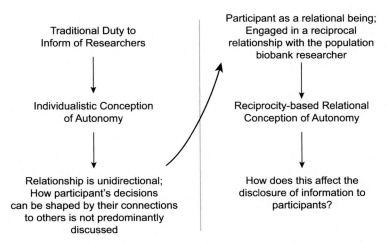

FIGURE 1

From an individualistic conception of autonomy to a reciprocity-based relational conception of autonomy.

generalization and offers an accessible point of entry for subsequent discussion. This does not mean, however, that the results of my research cannot be generalized and adapted to other fields in the future.

Second, and perhaps more importantly, I think that population biobanks best encapsulate the challenges facing the traditional way of understanding the consent process and the duty of researchers to inform in Canada. Indeed, both the recruitment of mostly asymptomatic participants and limitations on the initial provision of information by researchers challenge the current Canadian jurisprudential application of the duty to inform. In population biobanking, research participants are informed of the ultimate goal of the project in which they are enrolled: to improve the health of future generations and to benefit society at large. The emphasis put by population biobanks on stakeholders outside of the traditional researcher–participant relationship (such as society or the research community) will, given the currently prevailing conception of autonomy based solely on the individual participant, be a crucial element to consider.

BOOK STRUCTURE

In Chapter 1, I will demonstrate that the prevailing jurisprudential interpretation of the duty to inform in Canada is conceptually based on a theory of individualistic autonomy. To do so, I will first give an overview of the evolution of the duty to inform in Canada, both in the clinical and in the research contexts. In this account, I will trace the history of the evolution and describe how the physicians' duty to inform fundamentally shifted in the middle of the 20th century. As a consequence of this shift, clinical ethics moved away from paternalism, adopting a theory of individualist autonomy in its place. Further, I will argue that this shift had an outsized impact on fields of research and the duties of researchers. By explaining the current state of Canadian law, I will conclude Chapter 1 by showing how the dominant theme in contemporary research is individualistic autonomy. Finally, I will suggest that the influence of individualistic autonomy must be revisited in the context of population biobanking.

Chapter 2 focuses on describing the necessary characteristics of population biobanks as a way of differentiating them from other kinds of research. This assessment will reveal that the public and research communities play increasingly central roles in this kind of research. Using qualitative document analysis, I will review internal documents that Canadian population biobanks share with their participants with the goal of assessing what they have been promised at their time of enrolment. The findings of this chapter will, in turn, lead to an examination of the practical and theoretical limitations of the individualistic conception of autonomy in the population biobanking context.

Following the work undertaken in Chapter 2, Chapter 3 will focus on the practical limitations of the individualistic conception of autonomy. I will show that, despite the requirement that sufficient and adequate information be provided to population biobanking participants, the nature of population biobanking makes it challenging, as a practical matter, to provide such information to participants. More specifically, by drawing on the consent forms and associated documents reviewed in Chapter 2, in addition to policies, guidelines, and statements that have addressed the provision of information by researchers in population biobanks, I will demonstrate how population biobanks are constitutionally incapable of foreseeing every possible use of stored data and samples. This, as a matter of course, would entail that they must deviate from the requirement of full disclosure of all facts, probabilities, and opinions demanded by the individualistic conception of autonomy.

While Chapter 3 will discuss shortcomings of the individualistic conception of autonomy from a practical point of view, Chapter 4 will examine the matter from a more theoretical perspective. I will argue that individualistic autonomy is incapable of recognizing the complexities of benefit considerations in the research setting. Further, I will show that the individualistic conception of autonomy, with its unidirectional focus on participants, fails to make sense of the multilateral relationships that are necessarily implicated in population biobank research. This is especially true in the case of relationships involving the broader research community and the public at large. I conclude by outlining a number of solutions that have been proposed in the literature to address individual autonomy's shortcomings. Specifically, I will consider deliberative autonomy, principled autonomy, the duty to participate in research, and relational autonomy. Finding most of these solutions inadequate in the population biobanking context, I will reject the first three. I will point to relational autonomy, however, as a plausible basis for a conception of autonomy based on reciprocity.

Against this backdrop, I will turn to the concept of reciprocity in Chapter 5. First, I will provide a broad outline of the concept. Second, I will examine various proposed theories of reciprocity, explore potential reciprocal exchanges by outlining their nature, scope, flow, and overall value. Finally, I will demonstrate that the literature features two dominant conceptions of reciprocity: reciprocity for mutual benefit and reciprocity for mutual respect. Setting out this groundwork will allow me to adopt the concept of reciprocity as a basis for relational autonomy, thereby laying the foundation for a novel way of understanding the disclosure of information to participants in population biobanking. This will be the function of Chapter 6. Using the concept of reciprocity to identify the kinds of relationships that exist between stakeholders in the population biobanking context, I will demonstrate that reciprocity offers the most appropriate conceptual framework in which to situate relational autonomy. This is so largely because reciprocity-based relational autonomy is capable of acknowledging and sustaining the multilateral relationships implicated in population biobanking research without compromising the rights of research participants. I will present this argument by first giving an overview of the way in which reciprocity is conceived in existing biobanking literature. From there, I will identify the kinds of relationships that exist among the various stakeholders and how reciprocity provides a plausible conceptual mold for interpreting them. Finally, I will examine advantages and limitations of conceiving of reciprocity as a basis for relational autonomy in the way we approach the disclosure of information to participants during the consent process. I will do so by describing the observable characteristics of the reconceived standard of disclosure to participants that is externalized by the reciprocity-based relational conception of autonomy. More specifically, I will demonstrate how this new conception would allow researchers to conceive of participants as embedded within a web of relations and how they should not only be informed of the scope of their participation, but also of how their decisions may affect other stakeholders, including the public and the research community. I will finally conclude by considering future potential research on the topic.

References

[1] Dworkin RB. Getting what we should from doctors: rethinking patient autonomy and the doctor-patient relationship. Health Matrix 2003;13:235–9.
[2] Dworkin G. Paternalism. In: Zalta EN, editor. Stanford encyclopedia of philosophy; Winter 2017. Available from: plato.stanford.edu/entries/paternalism/.
[3] McCullough LB, Cross AW. Respect for autonomy and medical paternalism reconsidered. Theor Med 1985;6(3):295–308.
[4] Husak DN. Paternalism and autonomy. Philos Public Aff 1981:27–46.
[5] Gillon R. Paternalism and medical ethics. British Med J 1985;290(6486):1971.
[6] Childress JF, Mount Jr E. Who should decide? Paternalism in health care. London, England: SAGE Publications Sage UK; 1983.

[7] Weiss GB. Paternalism modernised. Journal of Med Ethics 1985;11(4):184—7.

[8] Laurie G. Genetic Privacy: a challenge to medico-legal norms. Cambridge: Cambridge University Press; 2002.

[9] Code of Ethics of Physicians. RRQ, c M—9, r 4. 1981.

[10] Buchanan A. Medical paternalism. Philos Public Aff 1978:377—82.

[11] Zawati MH. Liability and the legal duty to inform in research. In: Joly Y, Knoppers BM, editors. Routledge handbook of medical law and ethics. London: Routledge; 2015. p. 199—210.

[12] The National Commission for the protection of human subjects of biomedical and behavioral research, the belmont report: ethical principles and guidelines for the protection of human subjects of research. Washington, DC: US Government Printing Office; 1978. s 1 "Respect for Persons.".

[13] Philips-Nootens S, Lesage-Jarjoura P, Kouri RP. Éléments de responsabilité civile médicale: Le droit dans le quotidien de la médecine. 4 ed. Cowansville, Que: Yvon Blais; 2017.

[14] Robertson GB, Picard EI. Legal liability of doctors and hospitals in Canada. 5 ed. Toronto: Thomson Reuters Canada Limited; 2017.

[15] Halushka v. University of Saskatchewan. 53 DLR (2d) 436 at 443—444, 52 WWR (ns) 608 (Sask CA). 1965.

[16] Solomon WC. RJQ 731 at 743, 48 CCLT 280 (QCSC). 1989.

[17] Canadian Institutes of Health Research, Natural Sciences and Engineering Research Council of Canada & Social Sciences and Humanities Research Council of Canada. Tri-Council policy statement: ethical conduct for research involving humans. Ottawa: Secretariat Responsible for the Conduct of Research; 2014. art. 12.

[18] Knoppers BM, Zawati MH. Population biobanks and access. In: Canestrari S, Zatti P, editors. Il governo del corpo: Trattato di biodiritto, v. 2. Milan: Giuffrè Editore; 2011. p. 1181. Giuffrè Editore.

[19] Taylor K. Paternalism, participation and partnership—the evolution of patient centeredness in the consultation. Patient Edu Couns 2009;74(2):150—5.

[20] Pers TH, Karjalainen JM, Chan Y, Westra H-J, Wood AR, Yang J, et al. Biological interpretation of genome-wide association studies using predicted gene functions. Nat Commun 2015;6(1):1—9.

[21] Khoury MJ. The case for a global human genome epidemiology initiative. Nat Genet 2004;36(10):1027—8.

[22] CARTaGENE. Second wave information brochure for participants. 2014. https://cartagene.qc.ca/sites/default/files/documents/consent/cag_2e_vague_brochure_en_v3_7apr2014.pdf. [Accessed 21 March 2011].

[23] Master Z, Nelson E, Murdoch B, Caulfield T. Biobanks, consent and claims of consensus. Nat Methods 2012;9(9):885—8.

[24] Allen C, Joly Y, Moreno PG. Data sharing, biobanks and informed consent: a research paradox. McGill JL & Health 2013;7:85.

[25] Caulfield T. Biobanks and blanket consent: the proper place of the public good and public perception rationales. King's LJ 2007;18(2):209—26.

[26] Hansson MG. Striking a balance between personalised genetics and privacy protection from the perspective of GDPR. In: Slokenberga S, Tzortzatou O, Reichel J, editors. GDPR and Biobanking: Individual Rights, Public Interest and Research Regulation across Europe. Cham: Springer; 2021. p. 45—60.

[27] Shabani M, Chassang G, Marelli L. The Impact of the GDPR on the Governance of Biobank Research. In: GDPR and Biobanking: Individual Rights, Public Interest and Research Regulation across Europe. Cham: Springer; 2021. p. 31–44.

[28] Knoppers B, Abdul-Rahman MZ. Biobanks in the Literature. In: Bernice E, editor. Ethical Issues in Governing Biobanks: Global Perspectives. Farnham: Ashgate Publishing; 2008. p. 13.

[29] Zawati MH. There will be sharing: population biobanks, the duty to inform and the limitations of the individualistic conception of autonomy. Health LJ 2014;21:97.

[30] Caulfield T, Murdoch B. Genes, cells, and biobanks: yes, there's still a consent problem. PLoS Biol 2017;15(7):2–6.

[31] Beauchamp TL, Childress JF. Principles of biomedical ethics. 6 ed. Oxford: Oxford University Press; 2001.

[32] Merriam Webster Dictionary. Online edition, https://www.merriam-webster.com/dictionary/phenotype; [accessed 21.03.2011].

[33] Dove ES, Joly Y, Knoppers BM. Power to the people: a wiki-governance model for biobanks. Genome Biol 2012;13(5):1–8.

[34] Knoppers BM, Zawati MH, Kirby ES. Sampling populations of humans across the world: ELSI issues. Annu Rev Genomics Hum Genet 2012;13:395–413.

From paternalism to the individualistic conception of autonomy: a brief overview of the evolution of the medical duty to inform in the 20th century

1.1 Introduction

In this chapter, I will argue that the jurisprudential interpretation of the duty to inform of researchers in Canada is foundationally based on an individualistic conception of autonomy. In presenting this view, I will first give an overview of the evolution of the medical duty to inform and its underlying principles. In this context, I use the word "medical" to refer both to clinical and research settings. Although my primary focus is population biobanking—a research paradigm—I will begin by briefly examining the clinician's duty to inform. The reason for doing so is simple: the duty of researchers to inform has generally been interpreted in comparison to the duties of clinicians. For that reason, understanding the duty to inform in both contexts is necessary when examining the duty to inform in research. More specifically, this chapter will examine how the duty to inform in the clinical setting evolved in the second half of the 20th century. This, in turn, will help clarify how these changes have been effected in the research setting. I will then examine relevant Canadian case law and describe how it characterizes the duty of researchers to inform. Finally, I will outline the current conception of autonomy that is at the core of the duty of researchers to inform as it has been considered in Canadian case law.

Reciprocity in Population Biobanks. https://doi.org/10.1016/B978-0-323-91286-0.00008-3

1.2 From paternalism to the principle of respect for autonomy[1]

Respect for patient autonomy is a principle at the core of the medical duty to inform. Indeed, since the second half of the 20th century, patients have become central contributors to the therapeutic decision-making process. For centuries prior, however, a certain understanding of medical beneficence, as well as physician, pledges to protect patients from harm, justified widespread paternalistic practices [1,2]. As this chapter aims to give an overview of the evolution of the medical duty to inform, I will briefly examine the characteristics of paternalism, highlight its shortcomings, and describe how it waned over time. Further, I will outline autonomy's rise to prominence and describe how it became a guiding principle in medical practice and the basis of the medical duty to inform.

Contemporary scholars have defined paternalism as the "interference of a state or an individual with another person, against their will, and defended or motivated by a claim that the person interfered with will be better off or protected from harm" [3—6]. To act paternalistically, therefore, is to interfere with another's freedom of action, often on the presumption that doing so is for their own good. Paternalism, however, is not a monolithic concept. In order to understand the evolution of the duty to inform, it is useful to consider the identities of the "paternalist" actors in issue, as well as the class of persons with whom such paternalists interfere. In that sense, three distinctions may be made. First, paternalism may be narrow or broad in scope. Paternalism that is narrow in scope focuses only with state coercion [3]. Broad paternalism, on the other hand, is concerned with any paternalistic action stemming from the state, an institution, or private individuals [3,7]. A further distinction might be drawn between pure and impure paternalism. In pure paternalism, "the class being protected is identical with the class being interfered with" [7]. A classic example is of a physician who withholds information from a patient (ostensibly) for his or her own good. In impure paternalism, "the class of persons interfered with is larger than the class being protected" [7,8]. Dworkin, in his seminal essay on this topic, gives the example of a state that, recognizing potential harm to consumers, prohibits the manufacture and sale of cigarettes [9]. A third differentiation may be made between welfare and moral paternalism. Moral paternalism is typically associated with state intervention with the goal of protecting a person's moral well-being [10]. Welfare paternalism, on the other hand, aims at improving a person's quality of life [10]. Consider the 1847 *Code of Ethics of the American Medical Association*, which reads the following:

---

[1] Portions of this section have previously appeared in Zawati MH, There Will be Sharing: Population Biobanks, the Duty to Inform, and the Limitations of the Individualistic Conception of Autonomy. Health LJ. 2014; 21 and in Zawati MH, Knoppers BM, Population Biobanks and the Principle of Reciprocity. In: Hainaut P, Vaught J, Zatloukal K, Pasterk M. Biobanking of Human Biospecimens. Cham: Springer Nature; 2017. p. 99—109.

The obedience of a patient to the prescriptions of his physician should be prompt and implicit. He should never permit his own crude opinions as to their fitness, to influence his attention to them. A failure in one particular may render an otherwise judicious treatment dangerous, and even fatal [1].

This excerpt exemplifies pure welfare paternalism. It is pure because the class of persons protected (patients) is identical to the class being interfered with. It is welfare paternalism, moreover because interference aims at making the patient's life better.

In the medical context, physicians have long used pure welfare paternalism as a justification for withholding information from their patients [2,11—14]. Before criticizing this practice, Buchanan outlines the motivation of physicians for withholding information in the following way:

1. *The physician's duty—to which he is bound by the Oath of Hippocrates—is to prevent or at least to minimize harm to his patient.*
2. *Giving the patient information X will do great harm to him.*
3. *Therefore, it is permissible for the physician to withhold information X from the patient [5].*

This conclusion, (3), Buchanan writes, does not follow necessarily from the premises, (1) and (2) [5]. To demonstrate (3), Buchanan explains that an additional premise would be required. This additional premise would seek to assess whether providing a patient with information X would result in greater harm than withholding it [5]. This view would require that the physician exercise a comparative judgment [5]. The use of the word "judgment" in this context, moreover, implies something more than just a reflexive assessment. As a matter of practice, for a physician to withhold information X from a given patient, the required comparative judgment should "be founded on a profound knowledge of the most intimate details of the patient's life history, his characteristic ways of coping with personal crises […], and his attitude toward the completeness or incompleteness of his experience [5]." A judgment of this kind would almost certainly not be well founded if based solely on the abstract reasoning of the physician [5]. For this reason, such judgments would amount to what Dworkin characterizes as unjustified paternalism: an action that does not preserve or enhance an individual's ability "to rationally consider and carry out his own decisions [9]." As a consequence, the major shortcoming of pure welfare paternalism is its lack of respect for autonomy, which involves "attitudes and actions that ignore […] or are inattentive to others' rights of autonomous action" [15].

On its face, this is problematic given the fundamental role autonomy plays in many of our most important daily undertakings [15]. According to the Oxford English Dictionary, the word "autonomous" derives from the Greek words "auto" (self) and "nomos" (law), meaning "having one's own laws" [16]. Early use of the word autonomy did not refer to individuals, but to cities capable of enacting their own law [17,18]. When considered at the level of an individual, the word autonomy may refer to a variety of conditions, including the following: "the capacity of reason for moral self-determination" and the "liberty to follow one's will; control over one's own affairs; freedom from external influence, personal independence" [16]. Strictly speaking, autonomy requires at least two conditions: liberty and agency [15]. Accordingly, someone in a state of coma or another mental incapacity might not be considered autonomous.

Current interpretations of the respect for autonomy have been greatly influenced by philosophers Immanuel Kant and John Stuart Mill [19]. In his *Groundwork of the Metaphysics of Morals* [20], Kant claims that individuals have the capacity to determine their own moral destiny [20,21]. Based on the view that all human beings have unconditional worth, he argues that to violate a person's autonomy is to treat them as a means to an end, rather than as an end in themselves; "that is, in accordance with others' goals without regard to that person's own goals" [15]. As for John Stuart Mill, his essay "*On Liberty*" [22] focuses on the "individuality" of the autonomous individual. He asserts that only self-protection would warrant limiting an individual's liberty of action [22,23]. Otherwise, individuals should be allowed to pursue the lives they wish according to their own beliefs.

With the rise of the Western conception of individualism [24—26] and the mounting influence of the civil rights movement in the second half of the last century [27,28], paternalistic practices have declined and patient autonomy has emerged as an embodiment of personal freedom. The principle of respect for autonomy— crystallized by the doctrine of informed consent [18]—has become the foundational ethos in health care provision. This reality has shaped a positive duty for physicians to adequately inform their patients before and during the delivery of care. This duty is considered "positive" and involves "both respectful treatment in disclosing information and actions that foster autonomous decision making" [15].

Following the two seminal decisions by Canada's Supreme Court on the physician's duty to inform—*Hopp v Lepp* [29] and *Reibl v Hughes* [30]—there has been a keen focus on autonomy as a form of self-determination in Canadian law. Indeed, the Court in *Hopp v Lepp*, in its discussion of informed consent, states that the "underlying principle is the right of a patient to decide what, if anything, should be done with his body" [29,31]. In a similar way, when discussing the divulgence of risks in the informed consent process, Chief Justice Laskin, writing for the Court in *Reibl v Hughes*, alludes to the right of patients to know the risks of having or not having an operation or a treatment [30,32]. The Supreme Court of Canada has consistently taken this position. Indeed, in the *Ciarlariello* case, Justice Cory, on behalf of a unanimous Court, writes the following:

> *This concept of individual autonomy is fundamental to the common law and is the basis for the requirement that disclosure be made to a patient. If, during the course of a medical procedure a patient withdraws the consent to that procedure, then the doctors must halt the process. This duty to stop does no more than recognize every individual's basic right to make decisions concerning his or her own body [33].*

With that said, what are the legal characteristics of the medical duty to inform and what is its extent? Are there any limitations to the information that must be provided to patients or research participants? Section 1.3 of this chapter will discuss these issues by way of an analysis of pertinent Canadian case law addressing the nontherapeutic research setting. This discussion will highlight the particular conception of autonomy that underpins the legal requirements. However, as the duty to inform of researchers has been determined by Canadian courts through a comparison with the duty to inform in the clinical setting, I will also briefly explore the clinical context as well.

1.3 Medical duty to inform: characteristics in the clinical and the nontherapeutic research settings[2]

The essential character of the legal duty of physicians to inform in Canada is a requirement to provide patients with information sufficient to allow them to make the best possible decision when consenting to treatment. In *Hopp v Lepp*, the Supreme Court specified the scope of the physician's duty to inform, which they found to include a duty to answer:

Any specific questions posed by the patient as to the risks involved […] [and] […] without being questioned, disclose to [their patients] the nature of the proposed operation, its gravity, any material risks, and any special or unusual risks attendant upon the performance of the operation [29].

It is clear from this excerpt that the Court does not advocate "full disclosure" in the sense of a requirement that physicians disclose all risks to patients, no matter how remote. However, physicians *are* required to disclose any material, special or unusual risks, which, according to Chief Justice Laskin, are those that would carry significant consequences, even if such consequences are merely possible [29]. In *Reibl v Hughes* [27,34,35], Chief Justice Laskin introduces the "reasonable patient" standard when he writes that the duty to inform of physicians applies to what the physician knows or should know that his/her patient would deem relevant in making a decision about their care [30]. In successive case law, the requirements laid out in *Hopp v Lepp* and *Reibl v Hughes* have become a minimum standard with which physicians in common law provinces are expected to abide. In Quebec, risks must be disclosed to the patient if they are (1) probable and foreseeable; (2) rare, if serious and particular to the patient; (3) known to all, if particular to the patient; (4) important, if serious and decisive in the decision-making of the patient; and (5) increased, if a choice is possible [27]. With that said, Quebec civil law courts have tended to

[2] Portions of this section have previously appeared in Zawati MH, There Will be Sharing: Population Biobanks, the Duty to Inform, and the Limitations of the Individualistic Conception of Autonomy. Health LJ. 2014; 21.

reject the "reasonable patient" threshold proposed in *Reibl v Hughes* and have instead set out a test based on what a reasonable physician would disclose in the circumstances [36,37].

The duty to inform in nontherapeutic research contexts is higher in intensity than the duty that applies in the clinic. This difference in intensity originates in two leading decisions: *Halushka v University of Saskatchewan* and *Weiss v Solomon*. In *Halushka*, a student participated in an experiment on the use of a novel anesthetic and catheter insertion technique. The participant was informed that the procedure would last a couple of hours and was a "perfectly safe test" that had been "conducted many times before" [38]. During the procedure, the participant suffered a full cardiac arrest and remained unconscious for 4 days. Following the incident, the hospital withdrew the anesthetic from clinical use.

The participant survived and sued for damages. In its 1965 decision, the Saskatchewan Court of Appeal found that the disclosure of information that had taken place during the consent process had been inadequate. In its reasons, the Court contrasted the duty to inform in a research project with the equivalent duty in a clinical setting, writing that "the duty imposed upon those engaged in medical research [...] to those who offer themselves as subjects for experimentation, as the respondent did here, is at least as great as, if not greater than, the duty owed by the ordinary physician or surgeon to his patient" [39].

The Court then justified this heightened duty to inform by explaining that

> [t]here can be no exceptions to the ordinary requirements of disclosure in the case of research as there may well be in ordinary medical practice. **The researcher does not have to balance the probable effect of lack of treatment against the risk involved in the treatment itself.** The example of risks being properly hidden from a patient when it is important that he should not worry can have no application in the field of research. **The subject of medical experimentation is entitled to a full and frank disclosure of all the facts, probabilities, and opinions** which a reasonable man might be expected to consider before giving his consent [39]. (My emphasis)

Thus, the standard articulated by the Court is one in which lesser therapeutic benefit to a participant entails a correspondingly greater duty to inform. The Superior Court of Quebec reiterated this heightened duty to disclose in research in the 1988 *Weiss v Solomon* decision. In that case, a patient who underwent cataract surgery was invited to participate in a research project independent of the procedure. Over the course of the project, the participant was administered ophthalmologic drops and a fluorescein angiography. Following the fluorescein injection, the participant suffered a ventricular fibrillation and died [27,40]. The Court determined, among other things, that the risk of death or collapse due to the participant's preexisting heart condition had not been sufficiently disclosed. The Court, referring to *Halushka*, reiterated the importance of full disclosure in nontherapeutic research by characterizing the duty to inform in these contexts as the most exacting possible [41]. Put another way, the duty to inform in the research setting is more stringent than the disclosure requirements applicable in the clinic.

While these decisions reflect the present state of law on the duty to inform in research, I argue that the standard they set is undermined in an era in which observational health research is increasingly international, collaborative, longitudinal, and is less directly focused on individuals. More precisely, I argue that the full disclosure standard disproportionately focuses on research participants, while ignoring the place of other stakeholders embedded in the web of relations that exists in any given research project. In fact, nowhere in the two leading decisions is there a robust discussion of the responsibilities of participants toward other stakeholders in the research setting, nor is there consideration of the way that multilateral relationships in the research context might affect the standard of the duty to inform of researchers. This absence of clarity is exacerbated by the fact that the standard set by both the *Halushka* and the *Weiss* cases, developed as they were in consideration of clinical trials, is hardly generalizable. Indeed, contemporary health research features a diversity of methodological approaches that have not yet been considered by Canadian courts. Population biobanks, as we will see in Chapter 2, are a clear example of research projects that are, in terms of their nature and scope, quite different than those featured in the *Weiss* and *Halushka* decisions. Population biobanks constitute a compelling example of research initiatives that are longitudinal, collaborative, and interdependent on a number of stakeholders.

Before assessing the nature of biobanks in greater detail, it is important to understand the theoretical grounding of the decisions made in *Halushka* and *Weiss*. It is worth determining, in other words, whether there is a particular conception of autonomy that justifies the perception that participants are independent rather than in an interdependent relationship with other stakeholders. In order to better appreciate the duty to inform as portrayed by Canadian courts, it is first necessary to understand the conception of autonomy that this duty aims to respect. I will turn to this issue in the next section.

1.4 Origins of the conception of autonomy in *Halushka* and *Weiss*[3]

As I demonstrated above, the term "autonomy" captures a variety of concepts [18], including "the capacity of reason for moral self-determination" [16] and the "liberty to follow one's will; control over one's own affairs; freedom from external influence, personal independence" [16]. Indeed, autonomy is a concept broadly applied in the literature. It is often associated with "dignity, integrity, individuality […], responsibility, and self-knowledge" [42]. Owing to its relationship to this diversity of concepts, no single definition of autonomy emerges as uniquely authoritative. Gerald

[3] Portions of this section have previously appeared in Zawati MH, There Will be Sharing: Population Biobanks, the Duty to Inform, and the Limitations of the Individualistic Conception of Autonomy. Health LJ. 2014; 21.

Dworkin thus notes the following: "[w]hat is more likely is that there is no single conception of autonomy but that we have one concept and many conceptions of autonomy" [42].

"Individual autonomy" is often conceived as the most traditional conception of autonomy [17,43]. In the fields of bioethics [18,44] and medical law [3,33], this approach is widely applied—though certainly not without debate. Developing an understanding of individual autonomy may help to better contextualize the rationale supporting the *Halushka* and *Weiss* decisions and the requirement of full disclosure they establish in the nontherapeutic research setting, such as in observational research projects.

According to Onora O'Neill, individual autonomy: "[…] is generally seen as a matter of independence or at least as a capacity for independent decisions and action" [17]. The concept of individuality or "individual autonomy" can be traced back to John Stuart Mill's foundational work on utilitarianism. According to Mill, the rightful liberty of an individual can only be secured through the development of individual autonomy, which may only be interfered with in cases of self-protection:

> *That principle is that the sole end for which making is warranted, individually or collectively, interfering with the liberty of action of any of their number, is self-protection. That the only purpose for which power can be rightfully exercised over any member of a civilized community, against his will, is to prevent harm to others. He cannot rightfully be compelled to do or forbear because it will be better for him to do so, because it will make him happier, because, in the opinion of others, to do so would be wise, or even right [22].*

Mill's focus on individual autonomy stems from his belief that it ultimately forms one of the elements of well-being [22,45]. Roger Dworkin's writing may present a more concrete and contemporary understanding of individual autonomy, which he characterizes as the

> *[…] right of a patient to make his own decisions about important personal matters and to effectuate those decisions (or have them effectuated). Properly understood, this would mean that the **patient is entitled to all the information relevant to the decision**, including information the patient does not know he wants or needs. To exercise autonomy, the patient would have to be **fully informed** and counseled about what decision to make [3]. (My emphasis)*

Dworkin describes this conception as rooted in liberal individualism [3]. Similarities can be seen between Dworkin's proposal and the requirements set out in the *Halushka* and *Weiss* decisions. Indeed, the amplification of the duty to inform supported by the Canadian courts appears to be primarily motivated by liberal individualism. In *Halushka*, the court insisted that participants in nontherapeutic research have a right to a "full and frank disclosure of all the facts, opinions, and probabilities" raised by the research. This excerpt bears striking similarity to one of the explicit characteristics of liberal individualism, namely, the demand that a participant be "fully informed and counseled about what decision to make." In *Weiss*, moreover, researchers were expected to carry out full disclosure whether or not it was wanted by the participant [46]. Such an exacting disclosure requirement appears associated with liberal individualism, in which "the patient is entitled to all the information relevant to the decision, including information the patient does not know he wants or needs." Today, the individualistic conception of autonomy has inspired a standard in which "no waiver can be used to justify nondisclosure of information to a research subject" [27,35].

While an emphasis on individual autonomy—with its roots in liberal individualism—may help reduce paternalistic practices by physicians and researchers [47], it is not without significant shortcomings. Part of my argument aims to highlight such limitations in the research setting by using population biobanks as a case model. In order undertake this analysis, however, it will be important to understand the nature and characteristics of population biobank research, which will be outlined in the following Chapter.

1.5 Conclusion

In this chapter, I have aimed to explicate the dominant jurisprudential interpretation of the duty to inform of researchers in Canada. I described how courts have relied on a correlative understanding of autonomy as a way of supporting their assessment. As a way of understanding this way of reasoning, I traced the evolution of the 20th century duty to inform in Canada. Following this review, I concluded that while paternalism was once a widespread norm in both clinical care and research, respect for autonomy took its place as the basis of the duty to inform. From there, I demonstrated how researchers must conduct themselves in a way that respects an individualistic conception of autonomy when informing participants about their role in the research project. To be more precise, participants are seen as independent agents and not interdependent and situated within a web of relationships with other stakeholders. This state of affairs, I argued, has led to the adoption of an exacting duty to inform, one that requires researchers to fully disclose all facts, opinions, and probabilities when consenting participants for research. In later chapters, I will argue that such disclosure is impractical and, in some cases, even impossible. To do so, I will examine limitations of the individualistic conception of autonomy in the context of population biobanking. This will require that I first lay out the various essential characteristics of population biobanks and clearly differentiate them from alternative ways of conducting health research (Chapter 2). This characterization of population biobanks will later help in the development of a tangible understanding of the practical and theoretical limitations of the individualistic conception of autonomy.

References

[1] Chin JJ. Doctor-patient relationship: from medical paternalism to enhanced autonomy. Singap Med J 2002;43(3):152−5 (quoting Code of Ethics of the American Medical Association, 1847).

[2] Hinkley AE. Two rival understandings of autonomy, paternalism, and bioethical principlism. In: Engelhardt Jr HT, editor. Bioethics critically reconsidered. New York: Springer; 2012. p. 85−7.

[3] Dworkin RB. Getting what we should from doctors: rethinking patient autonomy and the doctor-patient relationship. Health Matrix 2003;13:235−9.

[4] McCoy M. Autonomy, consent, and medical paternalism: legal issues in medical intervention. J Alternat Compl Med 2008;14(6):785−92.

[5] Buchanan A. Medical paternalism. Philosophy & Public Affairs; 1978. p. 377−82.

[6] Rich BA. Medical paternalism v. Respect for patent autonomy: the more things change the more they remain the same. Mich St UJ Med & L 2006;10:87.

[7] Rodriguez-Osorio CA, Dominguez-Cherit G. Medical decision making: paternalism versus patient-centered (autonomous) care. Curr Opin Crit Care 2008;14(6):708−13.

[8] Tan N. Deconstructing paternalism: what serves the patient best. Singap Med J 2002; 43(3):148−51.

[9] Dworkin G. Paternalism. In: Wasserstrom RA, editor. Morality and the law. Belmont: Wadsworth Publishing Company; 1971. p. 183−8.

[10] Dworkin G. Paternalism. In: Zalta EN, editor. Stanford encyclopedia of philosophy; 2020. Winter ed 2017. Available from: plato.stanford.edu/entries/paternalism/.

[11] Weiss GB. Paternalism modernised. J Med Ethics 1985;11(4):184−7.

[12] Corn BW. Medical paternalism: who knows best? Lancet Oncol 2012;13(2):123−4.

[13] Gillon R. Paternalism and medical ethics. Br Med J 1985;290(6486):1971.

[14] McCullough LB, Cross AW. Respect for autonomy and medical paternalism reconsidered. Theor Med 1985;6(3):295−308.

[15] Beauchamp TL, Childress JF. Principles of biomedical ethics. 6 ed. USA: Oxford University Press; 2001.

[16] Oxford English Dictionary. Online edition. https://en.oxforddictionaries.com/definition/autonomous. (Accessed 21 03 11).

[17] O'Neill O. Autonomy and trust in bioethics. Cambridge: Cambridge University Press; 2002.

[18] Laurie G. Genetic privacy: a challenge to medico-legal norms. Cambridge University Press; 2002.

[19] Walker RL. Medical ethics needs a new view of autonomy. J Med Philos: A Forum Bioethics Philos Med 2008;33(5):594−608.

[20] Kant I. In: Hill Jr TE, Zweig A, editors. Groundwork for the metaphysics of morals. Oxford: Oxford University Press; 2009.

[21] Secker B. The appearance of Kant's deontology in contemporary Kantianism: concepts of patient autonomy in bioethics. J Med Philos 1999;24(1):43−66.

[22] Mill JS. In: Stillinger J, editor. Three essays. Oxford: Oxford University Press; 1975.

[23] Husak DN. Paternalism and autonomy. Philosophy & Public Affairs; 1981. p. 27−46.

[24] Childress JF, Mount Jr E. Who should decide? Paternalism in health care. SAGE. London: Publications Sage UK; 1983.

Table 2.1 Presentation of Canadian population biobanks.

Cohort name	Region(s) covered	Number of participants and age of recruitment	Purpose of the project	Access governance
BC Generations Project	British Columbia	29,848 [9] 35–69 years	This project seeks to "help researchers learn more about how environment, lifestyle, and genes contribute to cancer and other chronic diseases." [10]	Access Committee (controlled-access); also part of the CanPath National Access Process
The Tomorrow Project	Alberta	54,184 [11] 35–69 years	This project seeks to "understand what causes diseases such as cancer, heart disease, and other long-term health conditions." [12]	Access Review Panel (controlled-access); also part of the CanPath National Access Process
CARTaGENE	Quebec	43,000 [13] 40–69 years	"CARTaGENE project will help to provide a better understanding of how our environment, lifestyle, and genetic background inherited from our parents are involved in the development of chronic diseases such as diabetes, cancer, and heart disease. This could improve the prevention, diagnosis, and treatment of diseases, and therefore, contribute to the improvement of the Quebec health system." [13]	Data and Sample Access Committee—SDAC (controlled-access); also part of the CanPath National Access Process
Ontario Health Study	Ontario	225,000 [14] 18 years and older	This project seeks to investigate "risk factors that cause diseases like cancer, diabetes, heart disease, asthma, and Alzheimer's." [15]	Data Access Committee (controlled-access); also part of the CanPath National Access Process
Atlantic PATH	Prince Edward Island; New Brunswick; NL; Nova Scotia	35,935 [16] 35–69 years [16]	This project seeks to "help researchers find out why some people develop cancer and others do not, so that we can find new ways of	Data Access Committee (controlled access); also part of the CanPath National Access Process

Continued

Table 2.1 Presentation of Canadian population biobanks.—*cont'd*

Cohort name	Region(s) covered	Number of participants and age of recruitment	Purpose of the project	Access governance
Canadian Alliance for Healthy Hearts and Minds	British Colombia; Alberta; Ontario; Quebec; Prince Edward Island; New Brunswick; Newfoundland and Labrador; Nova Scotia [17]	9700 [18] 35–69 years	preventing this disease. It will also help us find new ways to diagnose cancer earlier, when it can be easier to treat" [17]. This project has 3 principal research objectives: (1) "To understand the role of socio-environmental contextual factors on individual risk factors, subclinical disease, and events. (2) To identify unique patterns of contextual factors, risk, health service utilization, and clinical outcomes in high-risk groups including Aboriginal people, Asian, Afro-Canadians. (3) To identify markers of early subclinical dysfunction of the brain and the heart and describe their relationship to individual/contextual risk, and outcome" [19].	Alliance Data Access Committee For non-CanPath participants; CanPath Access Committee for CanPath participants (controlled-access)
Canadian Longitudinal Study on Aging	British Colombia; Alberta; Manitoba; Ontario; Quebec; Nova Scotia; NL [20]	51,352 [21] 45–85 years	This project seeks "to find ways to improve the health of Canadians by better understanding the aging process and the factors that shape the way we age" [21]. It examines healthy aging by studying the changing biological, medical, psychological, social, lifestyle, and economic aspects of people's lives.	Data and Sample Access Committee—DSAC (controlled-access) [22]

this range allows researchers to better observe the progression of disease among participants. This is so simply because the period is both sufficiently large and late in life to see participants begin to fall ill [2,23]. Developing an understanding of illnesses observed in population biobank participants requires that data and samples are collected over time. In addition, research results will, where possible, be linked with personal health data provided by administrative databases. To better understand these features, the following sections will outline the following characteristics of population biobanks: (1) that they are essentially *for* the public; (2) that they are established to supply data and samples for future research projects; (3) that they are linked with administrative health data; and, finally, (4) that they are organized and searchable collections.

2.3 Project of the public, by the public, for the public

The holding of a population-based collection is perhaps the most unique feature of population biobanks. It is a primary distinction between them and other kinds of data and sample repositories. The rationale behind the establishment of biobank collections stems from the aim of studying common, complex diseases that are prevalent in any given population [2,23,24]. "Complex" in this context refers to diseases that are multifactorial in nature. While researchers are learning that nearly every disease has some genetic component, many, such as heart conditions or obesity, are believed to be associated with multiple gene interactions in addition to environmental and lifestyle considerations [25]. For most of the history of medicine, the way these factors contributed to disease was not well understood [26]. Consequently, there was insufficient knowledge to fully understand these diseases and positively impact public health initiatives or patient care [26]. The translation of genomic discoveries into the clinical setting promises to change that reality. With that in mind, genomics has motivated a number of countries to establish large-scale population-based studies [26,27] that aim to link biomarkers to medical history and lifestyle information [26].

Examples of national population studies include the Canadian Longitudinal Study on Aging (50,000 participants aged 45−85) [21] and the UK Biobank (500,000 participants aged 40−69 years) [28]. In both projects, individuals representative of the general population are randomly selected and asked to participate. Invited participants are asked to appear at various assessment centers, where they provide certain data and samples [5]. Other population biobanks recruit participants through clinicians. This is the approach taken, for example, by the Estonian Biobank (51,535 adults aged 18 years and older) [29] and the Lifelines Project in the Netherlands (165,000 participants, across three generations) [30]. After giving consent to participate in the study, these mainly asymptomatic individuals are asked to provide biological samples and data derived from self-administered and interviewer-assisted questionnaires. The fact that most participants are asymptomatic is quite important, as it indicates that they should not expect to obtain any direct therapeutic benefit from their participation.

Finally, population-based initiatives are typically not limited to a single jurisdiction. In recent years, infrastructures networking population biobanks from different geographical locations have begun to emerge. Examples include the Biobanking and Biomolecular Resources Research Infrastructure ("BBMRI") (pan-European, 53-member consortium) [31], the Canadian Alliance for Healthy Hearts and Minds and the Canadian Partnership for Tomorrow's Health project ("CanPath") (a pan-Canadian research study of five population cohorts: BC Generations Project, Alberta Tomorrow Project, Quebec's CARTaGENE, Ontario Health Study, and the Atlantic PATH project). These networks explore how genetics, environment, lifestyle, and behavior contribute to the development of cancer and other chronic diseases [32]. Such collaborative endeavors increase the statistical power of the overall collection [23,33] and facilitate related ethical, legal, and social issues policy interoperability [34].

2.4 Established to supply data and samples for future research projects

This second characteristic of population biobanks is that they are established to supply data and samples for future research projects. This feature is shared by all research biobanks, whether they are disease specific [35] or composed of residual samples collected following medical care [36]. In fact, the very goal of instituting a research biobank is to supply searchable data and samples for future research projects [37–39]. What distinguishes population biobanks, however, is the kind of consent procedure applied during recruitment [40]. This different approach, which I will explain below, is justified in light of two realities: (1) the supply of data and samples in population-based biobanks is generally more frequent given the size of the collection [41] and (2) access by outside researchers for presently unspecified projects may occur many years in the future [42]. These two points are reflective of a critical reality: that stakeholders outside of the participant–researcher paradigm play an essential role in the success of population biobanks. The results I describe in this section demonstrate the important role played by the research community on the issue of data and sample access in the Canadian setting. To my knowledge, no detailed review of the way population biobank participants are informed of how researchers will access their data and samples has yet been performed from a Canadian perspective.

2.4.1 Methodology

In this section, I sought to develop an understanding of how large-scale Canadian population biobanks have approached the future use of data and samples. My objective was to highlight any role played by stakeholders other than research participants. Elucidating a role of this kind would help underscore whether truly research participants are interdependent agents embedded in a web of relations. In applying this methodology, I analyzed consent forms, information brochures, and Frequently Asked Questions (FAQs) posted on population biobank websites. These documents reflect the extent of information provided to participants during the recruitment process. Understanding what population biobanks present to their participants when obtaining informed consent allows for an assessment of the ease or difficulty these projects have in actually disclosing information about data and sample access to participants.

No search engine was useful in the identification of Canadian population biobanks. Therefore, I relied on working knowledge of existing biobanks to provide guidance in this search. This allowed for the identification of population biobanks from various provinces that I have studied in previous work on this topic. More specifically, I made use of document analysis, a qualitative research methodology, to both identify these documents and analyze their content. This method is defined as a "systematic procedure for reviewing or evaluating documents—both printed and electronic" [43]. Document analysis is considered an analytical method in qualitative research, where data are "examined and interpreted in order to elicit meaning, gain understanding, and develop empirical knowledge" [43]. More precisely, the analytical procedure involves finding, selecting, understanding, and synthesizing data found in documents. Containing elements of both content analysis and thematic analysis, document analysis "entails a first-pass document review, in which meaningful and relevant passages of text or other data are identified" [43].

I identified a total of 22 documents from seven biobank projects. Some of these were obtained online, such as those from CARTaGENE, the Canadian Longitudinal Study on Aging, and the FAQs of the Ontario Health Study. Documents from BC Generations, the Alberta Tomorrow Project, Atlantic PATH, and the Canadian Alliance for Healthy Hearts and Minds were received through correspondence with the scientific directors of each project. The Canadian Alliance for Healthy Hearts and Minds provides consent forms from 14 different sites in Canada. 3 sites were chosen randomly for inclusion (Aboriginal Participants Site, Montreal Heart Institute, and Thunder Bay). These sites are generally representative as the documents from all 14 sites contain nearly identical information.

Not using a particular search engine and relying on working knowledge to identify population biobanks could create selection bias. For example, new population biobanks or those of which I am unaware may not have been included. Thanks to my role as Access Officer for the CanPath project, previously known as the Canadian Partnership for Tomorrow Project, I have tried to palliate this limitation by staying abreast of new and emerging biobank projects. The Manitoba Tomorrow Project is a case in point. This new population biobank only began to enroll research participants in 2017−18 [44]. After reviewing this project's consent form, I decided not to include the project as the content of the form added no new information to what I had already collected. Indeed, being the biobank to most recently join the CanPath project, the Manitoba Tomorrow Project has drawn heavily on the consent forms of other cohorts within the consortium (CARTaGENE, AtlanticPATH, BC Generations, Alberta Tomorrow Project, and Ontario Health Study). Furthermore, population biobanks of which I am less familiar, such as the Canadian Longitudinal Study on Aging, have been studied to ensure as little selection bias as possible.

In conducting the document analysis, I first screened the 22 included documents for pertinence—that is, whether information was related to that provided during the recruitment of participants and the administration of consent. The scope of information provided to participants at that moment, and captured in documents and brochures, provides a tangible way to assess the nature and limitations of the duty to inform of researchers working in the population biobanking context. Documents not directed at participants and therefore not part of the consent process (e.g., access policies, access agreements) were excluded (12). Such documents are generally intended for internal staff members or outside researchers. They normally contain technical information. The remaining 10 documents were thoroughly analyzed to identify approaches used and mechanisms for the future use of data and samples. Given that my primary objective was to highlight the existence of any potential role for stakeholders other than research participants, examining these documents will be helpful in that regard. Indeed, document analysis was used to analyze selected documents. More precisely, a theory-driven approach (used when there is already knowledge of the themes) was used to advance the general theme associated with the future use of data and samples. For this theme, subcategories were identified and coded; for example, the subcategories "use," "access," "data and samples," and "request" were identified for the theme "access." This method allowed for the identification of common patterns related to the future use of data and samples across a number of consent forms, information brochures, and FAQs posted on population biobank websites. Pertinent passages containing information retrieved through this method were highlighted and organized in a table (Table 2.2).

According to Glenn Bowen, the document analysis research methodology has several advantages. These include efficiency, availability of documents, cost effectiveness, stability (analyzed documents are not altered), a lack of obstructiveness, and reactivity (documents are unaffected by the research process) [43]. Given that my only purpose was to review and analyze the content of consent forms and similar documentation, rather than understand how they were interpreted by research participants or the experiences of researchers in administering them, document analysis proved more pertinent than, say, interviews or surveys might have been. There are, however, some limitations presented by the chosen methodology. These include insufficient detail (some documents might not provide sufficient detail to answer a research question), low retrievability (some documents are difficult to access), and biased selectivity (the available documents are likely to be aligned with the agenda of the organization that adopted them) [43]. These limitations were constrained in my analysis, given that the documents selected provided sufficient detail to answer my research question and were, for the most part, retrievable. Those that were not publicly available were made accessible via correspondence. As for biased selectivity, this limitation would likely have a stronger effect in an organizational context, in which one is analyzing the internal policies of organizations (such as human resources documents) [43]. In my case, the fact that the documents are aligned with the

Table 2.2 Consent provisions addressing access to data and samples from Canadian population biobanks.

Name of the cohort	Portion of the cohort documentation
BC Generations Project *(British Columbia)*	**Consent form (version 4.0—December 12, 2014)** "Your information and samples will be collected, coded, and stored at highly secure and protected sites at the Cancer Research Centre of the BC Cancer Agency." "The BC Generations Project expects to receive requests from Canadian and overseas scientists and international collaborators to use the information or your sample (with your identifying information removed). All access will be subject to the strictest scientific and ethical scrutiny and independent oversight" [45].
The Tomorrow Project *(Alberta)*	**Consent form (DS-3008Av3 CPTP (Now CanPath) combined consent—May 2011)** "I accept that my data and samples may be used, in coded form, by approved researchers from Canada and other countries for research related to cancer, and potentially other health conditions, and this will continue even after my death or if I can no longer make decisions" [46]. **Study Booklet (version DS3010v2—May 2011)** "Researchers may apply to access the research data and samples that are stored by the Tomorrow Project in Alberta." [47] "Applications for access to data or samples may be received from, and approved for, researchers working in Alberta, other parts of Canada, or international locations." [47]
Ontario Health Study *(Ontario)*	**Consent form (version 10—April 24, 2014)** "I accept that my information and blood sample, after my name and other identifying information have been removed, may be used by researchers from Ontario, Canada (e.g., as part of the CanPath project), and other countries for approved health-related research projects" [48]. **OHS website FAQ** "All data and information that you provide will be kept on secure servers at the Ontario Institute for Cancer Research, housed in Toronto, Ontario" [49].
CARTaGENE *(Quebec)*	**Information brochure with consent form (April 7, 2014)** "CARTaGENE will only grant access to data and samples to authorized researchers. Access will not be authorized to insurance companies and employers. [...] As so, the data and samples will be coded. [...] The data and samples collected for the CARTaGENE project will be used for research on health and/or genomics. Researchers with projects that have been approved can ask to use certain samples and data. In this case, ethics committees will evaluate the research projects submitted and the scientific validity of these studies will be examined by an access committee independent from CARTaGENE [50].

Continued

Table 2.2 Consent provisions addressing access to data and samples from Canadian population biobanks.—*cont'd*

Name of the cohort	Portion of the cohort documentation
Atlantic PATH *(Atlantic Provinces)*	**Consent and brochure (version 9.2—March 6, 2013)** "We expect to receive requests and, if approved, provide Canadian and International Researchers access to the data and samples. A Research Ethics Board, like the one that helps protect you during this research project, will review and approve all future projects before other researchers gain access to your samples. We may share the samples with other researchers, but we will not give the researchers any information that would allow them to identify you. We will always know which sample belongs to you, but other researchers will not" [51].
Canadian Longitudinal Study on Aging *(Canada)*	**Study information package—home interview and data collection site visit** "The CLSA Data and Sample Access Committee must approve requests from researchers from Canada and other countries to use your data and samples" [52]. **Consent form—home interview and data collection site visit** "I understand that my information and samples will be used for research purposes only and this research may also have commercial uses that benefit society" [53].
Canadian Alliance for Healthy Hearts and Minds—Aboriginal Participants	**CAHHM—participant information and consent sheet (aboriginal participants)** "[…] qualified national and international researchers will be able to access to it for future research projects" [54].

intentions of their developers was precisely the point. Beyond that, data triangulation with the information retrieved from the literature [55], as well as other sources, allowed me to establish consistency and to corroborate my findings [43].

2.4.2 Results

Given the limited amount of specific information available during the recruitment phase, most population biobanks have resorted to what is commonly referred to as "broad consent" [56] (discussed at greater length in Chapter 3). The term "broad consent" or "general consent" means "consenting to a framework for future research of certain types" [57] and pertains "to a bank or research infrastructure whose possible uses are not all known at the start" [58]. This approach contrasts with "specific consent," in which participants give consent for the use of their data and samples in a given area of research or disease type for a limited period of time [59]. Indeed, in cases of specific consent, future use that does not fall within the definitive parameters described in the consent form demands that biobanks reconsent their participants for the relevant secondary use [60]. This is certainly not the case for Canadian population studies, as is evidenced in the clauses included in Table 2.2 below.

While some variability exists, four common themes can be drawn from the above selected clauses: (1) jurisdiction of applicants; (2) type of data/samples being provided; (3) scope of the projects undertaken by applicants; and (4) bodies adjudicating access requests. In fact, in all of these examples, research participants are informed during recruitment that future applicants for data and sample access may be either Canadian or international researchers. Participants are also informed whether any restrictions on access will be made based on the national status of researchers. Some Canadian biobanks have imposed restrictions on access by insurance companies and employers [50]. In all of the examples given above, the population biobanks also specify the type of data and samples that will be supplied for future research projects. Much of the time, these data are coded [61], meaning that "direct identifiers are removed from the information and replaced with a code" [60]. Coding reduces the risk of a breach of confidentiality by outside researchers and allows biobank operators to reidentify the participant, if necessary, or to link their information with administrative health data (discussed in Section 2.5). Indeed, if identifiable information is irreversibly removed, the data and samples cannot reasonably be linked back to the research participant in question [62]. The Global Alliance for Genomic and Health's *Privacy and Security Policy*, for example, defines anonymized data as "data that were related to an identifiable individual when collected, but through a process of removing all direct identifiers, thereafter prevents the identity of an individual from being readily determined by a reasonably foreseeable method" [62].

A third theme relates to the scope of research domains for which data and sample access will be permitted. Consent forms with more encompassing research domains, for example, allow wider access to data and samples. This, however, certainly does not entail a blank check for researchers to undertake any kind of research in any given field and framing will be necessary to ensure that participants who agree to future use are not providing blanket consent. Such framing is clearly evidenced in the consent form clauses above. Population biobanks in Canada, when describing the type of projects that can be undertaken with their data and samples, refer to "health-related research projects" [48], "research on health and/or genomics" [50], "research related to cancer and potentially other health conditions" [46]. A more specific consent approach would require that either the population biobank pinpoint an exact project or disease that would have use of relevant data and samples [63] or to frequently recontact participants to renew consent every time a new project requests access to the repository. These strict parameters would subsequently limit access to the resource and are, by and large, impracticable [59].

Finally, it is evident in Table 2.2 that consent forms generally refer either to entities such as those that adjudicate requests for access, such as the "CLSA Data and Sample Access Committee" [52] or to "access committee independent from CAR-TaGENE" [50]. I will describe the governance surrounding access mediated by these entities in greater detail in Section 2.6 of this chapter. For the moment, I will turn to the third characteristic of population biobanks that of their linking research data with administrative health data.

2.5 Linked with administrative health data

In order for researchers to better understand the multifactorial nature of common, complex diseases, there is a need to link biomarkers with medical history and lifestyle information [40]. This is the third foundational characteristic of population biobanks. Such linkage makes high-quality data more readily available, "including data on individuals and their encounters with service providers in the health system as well as social data on factors that affect health outcomes" [64].

Defined as the bringing together of data relating to the same individual from two or more sources [65], data linkage involves the use of a common identifier "such as personal health number, date of birth, place of residence, or sex" [65] to combine data related to the same individual available in other databases. In population-based biobanks, there are a number of reasons why such linkage is instrumental [66]. Chief among them is the ability to enrich "study datasets with additional data not being collected directly from study participants" [65]. Such linkage offers "vital information on health outcomes of participants, and serve to validate self-reported information" [65,67]. Indeed, using "additional data which records such information as a matter of course can improve the accuracy of data collection and reduce the burden on both observer and subject" [66].

2.5.1 Methodology

In this section, I used the same consent forms, information brochures, and FAQs analyzed in the previous section. As before, I take document analysis as my methodological approach, though I add one additional caveat. At this stage of research, the remaining 10 documents were reviewed in order to better understand how large-scale population biobank studies in Canada have dealt with linkage to administrative health databases. A deductive thematic approach was again taken to advance the general theme of access to administrative health data. Here, as above, subcategories were identified. More precisely, the subcategories "access," "health services," "registry," "administrative health databases," and "records" were identified for the theme "access to health administrative data." Following this, the documents were coded. This method allowed for the identification of common patterns across consent forms, information brochures, and FAQs posted on population biobank websites. Pertinent passages containing information retrieved through this method were highlighted and organized into a table (Table 2.3). Consideration of linkage issues is critical, for it is not only an essential characteristic of population biobanks, but also reinforces the need to cross-check collected data with administrative records as a way of ensuring accuracy and correlatively accelerating the proliferation of public health benefits.

2.5.2 Results

As evident in the consent clauses presented in Table 2.3, databases used for linkage purposes include, but are not limited to the following: cancer registries, health, and

Table 2.3 Consent provisions addressing data linkage from Canadian population biobanks.

Name of the cohort	Portion of the cohort documentation
BC Generations Project *(British Columbia)*	**Consent form (version 4.0—December 12, 2014)** "We are asking your permission to access information on your health and health procedures that may occur in the future, or may have occurred in the past, as far back as 1985. The sources of this information include existing electronic data files such as the following: **BC Cancer Agency:** The BC Cancer Agency keeps a highly confidential and accurate registry of all cancer cases diagnosed in British Columbia, and all deaths from cancer in the province as well as information on screening procedures and cancer treatment. Information from you will be linked to the BC Cancer Agency databases. **Population data BC:** The BC Ministry of Health keeps confidential records of the health services used by all residents, and these records are the most accurate and complete source of this type of information in British Columbia. A study about the causes of disease needs to include information about chronic diseases developed as well as the types of health care services people need, how often services are used, and whether the services are provided at a doctor's office or in a hospital." [45]
The Tomorrow Project *(Alberta)*	**Consent form (DS-3008Av3 CPTP (now CanPath) combined consent—May 2011)** "I accept that the Tomorrow Project may request additional information from health records and databases (including, but not limited to Alberta Cancer Registry and Alberta Health and Wellness databases) about my past, current and future health, and will continue to do so even if I can no longer make decisions or after my death." [46] **Study Booklet (version DS3010v2—May 2011)** "We are asking your permission to access past, current, and future health records and administrative health databases. […] Health records and databases can also help explain how patterns of health services used over time may be associated with long-term health. Examples of databases that may be accessed by the *Tomorrow Project* include the following: **Alberta Cancer Registry**. The Alberta Cancer Registry is legally responsible for keeping an accurate record of all cancer cases diagnosed in Alberta, and all deaths from cancer in the province. […] The *Tomorrow Project* will need to know the type of cancer, when it was diagnosed, what the diagnostic stage was, and if it was a particular subtype defined by a special laboratory test.

Continued

Table 2.3 Consent provisions addressing data linkage from Canadian population biobanks.—*cont'd*

Name of the cohort	Portion of the cohort documentation
	Alberta Health and Wellness Databases. Alberta's provincial health ministry keeps information on the health services used by Alberta residents. […] For example, this database could be used to tell us which participants have had colorectal cancer screening tests, and when. This kind of information could be important in understanding how the use of colorectal cancer screening tests affects the numbers of people who develop this kind of cancer" [47].
Ontario Health Study *(Ontario)*	**Consent form (version 10—April 24, 2014)** "I understand that the information and samples I provide will be linked with information about me found in both current and future health-related databases (e.g., Ontario Health Insurance Plan (OHIP) Claims Database, Ontario Cancer Registry), in my personal medical records, and with any additional information I might provide in the future" [48]. **OHS website FAQ** "[…] For example, every time you undergo certain tests (e.g., a mammogram), the fact that you had this test is noted and stored in a database. This is referred to as "administrative data." By linking the information you provide to the OHS with administrative data, researchers are able to ask a broader range of questions, such as whether screening programs are effective and whether there are "hot spots" across the province where a certain disease is more common" [49].
CARTaGENE *(Quebec)*	**Information brochure with consent form (April 7, 2014)** "I accept that personal information about me contained in government health administrative databases be transmitted confidentially to CARTaGENE in coded form when needed for research in health and genomics. This information may cover the period from January 1st, 1998 to the end of the CARTaGENE project" [50].
Atlantic PATH *(Atlantic Provinces)*	**Consent and brochure (version 9.2—March 6, 2013)** "If you agree to participate, you will also be allowing us permission to access routinely collected information on health procedures you may undergo or may have undergone in the past. The sources of this information include existing electronic data files such as the following: Cancer Care Nova Scotia is responsible for keeping a highly confidential and accurate registry of all cancer cases diagnosed in Nova Scotia. This information is used to estimate the rates of new and existing cancer in the population and death rates from various types of cancer. […] The PATH study will also be accessing Vital Statistics records related to death records. The Nova Scotia Department of Health keeps confidential

Table 2.3 Consent provisions addressing data linkage from Canadian population biobanks.—*cont'd*

Name of the cohort	Portion of the cohort documentation
	records of the health services used by all residents, and these records are the most complete source of this type of information in Nova Scotia. A study about the causes of disease needs to include the types of health care services people need, how often services are used, and whether the services are provided at a doctor's office or in a hospital" [51].
Canadian Longitudinal Study on Aging (*Canada*)	**Study information package—home interview and data collection site visit** "Your provincial health care records will be linked to data collected by the CLSA to study patterns of health and health care over time. For example, Ministries of Health in each province keep records about your visits to doctors and hospitals, medicines you fill a prescription for, and what people die from" [52]. **Consent form—home interview and data collection site visit** "I understand that if I choose to give my Health card Number, it will be used to link information about me in my public health care records held by the Provincial Government" [53].
Canadian Alliance for Healthy Hearts and Minds—Montreal Health Institute (MHI) Site	**CAHHM—participant information and consent sheet (MHI site)** "When you agreed to participate in the MHI Biobank, you may have provided your health card number so that your study file could be linked with the Quebec health insurance system (RAMQ) database. This allows us to obtain additional information on your long-term health status by accessing information directly from the RAMQ database, for example, and merging it with your Alliance participant file" [68].

wellness databases held by governmental entities, and bodies curating personal medical records of patients. In Canada, linkage to administrative health databases is regulated by provincial authorities from whom approval must be sought—even when participants have consented [65]. For national projects with multiple sites across the country, this provincial fragmentation tends to impede timely access by researchers interested in obtaining nationally representative data—a matter that has prompted several deliberations and initiatives aimed at creating a unified national framework [64]. Exploring these endeavors in greater detail, however, is beyond the scope of this book.

Table 2.3 outlines the importance placed on data linkage by population biobanks. More precisely, it indicates the multiple sources of such linkage in the provinces and corroborates the level of emphasis provided in the literature on the significance of these data.

Following a review of these clauses, several observations can be made. First, all administrative health databases mentioned in the consent forms are provincial, as pointed out by the literature referenced in this section. If a project is hosted in British Columbia, for example, only data stored in governmental/administrative databases in that province will be accessible. Even pan-Canadian projects, such as the Canadian Longitudinal Study on Aging, specify the provincial nature of agencies and ministries for linkage purposes. Given that most of these research projects study cancer and other chronic diseases, three main sources of data are mentioned in most of the consent forms: (1) ministries of health, (2) cancer registries, and (3) health services databases. Second, and perhaps more interestingly, information surrounding linkage takes up more space than any other section in the consent forms of a majority of analyzed documents (7/10). This is a testament to the importance of these procedures for the biobanks. Finally, most projects explicitly express the rationale supporting linkage by informing participants that linkage procedures "help explain how patterns of health services used over time may be associated with long-term health," "researchers are able to ask a broader range of questions," and linkage allows the biobank to "obtain additional information on your long-term health status." While participant biobanks are observational in nature, linkage procedures provide limited researcher contact with the clinical setting, in the sense that most of the information found in the government health databases will be clinical in nature. Having access to such information and cross-checking it with self-administered data collected by biobanks allows them to verify and "clean" the relevant data. Doing so permits the use of only validated information in the translation of knowledge from the research setting to the clinic. I will now turn to the fourth and final relevant characteristic of population biobanks: the organized and searchable nature of their collection.

2.6 Organized and searchable collection

The fourth characteristic of population biobanks is the organized and searchable nature of their collections. Organization is a central characteristic of *any* research biobank, but the practices of population-based studies differ relative to the nature of the collection, the frequency of access requests to their data and samples, their longitudinal nature, and the level of communication projects have with their participants. Organization in a population biobank is guided by both internal and external governance. More specifically, population biobanks create governance mechanisms that ensure oversight, management, access, use, and closure of the biobank, communication with participants, and compliance with legal and ethical principles [69]. Processes are put in place to review, update, and modify governance policies over time [70]. The overarching goal of each of these initiatives is, ultimately, to sustain public trust [24,71]. Indeed, "it is not enough to ask a whole population for unquestioning trust, one must put in place good governance and mechanisms to ensure that the projects follow through with their promises to participants" [72]. As I mentioned above, governance mechanisms take multiple forms. For the purposes of this text, I only focus on two internal governance mechanisms, which will then be used in upcoming sections to frame discussion of issues raised by the individualistic conception of autonomy.

The first common governance mechanism relates to operations management. To keep an organized and searchable collection, biobanks implement mechanisms to establish and oversee standard operating procedures (SOPs), quality control, and quality assurance, among other things [73]. SOPs, for example, are important for standardizing the preparation and storage of data and samples. They may also be used to ensure consistency in a project that involves activities taking place at different sites. Operations management further includes the establishment of various committees mandated to lead certain areas of the biobank's activities [22,74]. For example, an Operations Steering Committee will be created to ensure scientific leadership of a study and to help determine and shape the milestones of the biobank throughout its term [22]. An ethics and legal committee may be instituted to oversee the development of policies concerning privacy, the return of research results and incidental findings, access (which I will discuss in greater detail below), publications, and intellectual property [22,75]. This kind of committee generally acts in an advisory capacity and assists in the development of consent forms for the biobank. The advantage of having an ethics and legal committee in large-scale biobanking is that issues related to public engagement, legal and ethical compliance, and data protection in legislation are handled by experts in these fields. If the population biobank spans multiple jurisdictions, the ethics and legal committee could be tasked with analyzing the legislative landscape across the different regions in order to develop a more harmonized approach.

One essential component of operations management is the facilitation of communication with the public and participants. Communications with the public might include the publication of a website that provides information on the project and its milestones [76]. Other public communications might include the organization of citizen forums [77] and deliberative engagement sessions [78,79]. As for communication with the participants, the publication of newsletters [80] and formal recontact procedures, provided the participants have consented to such contact [81,82], may satisfy that goal.

A second governance mechanism concerns access to data and samples. In order to sustain public trust in population biobanks, the implementation of an ethical, economic, and efficient access system is of fundamental importance. Doing so involves not only the development of required documentation but also the creation of bodies tasked with evaluating and approving access requests [83]. In essence, biobank participants have agreed to have their data and samples used in future, yet unspecified research projects. This necessitates mechanisms for ensuring that the process is carried out in a manner that respects the wishes of participants as expressed in their consent forms and protects their privacy and the confidentiality of their data and samples [84]. Documents created for these purposes generally include an Access Policy, Publications, and Intellectual Property Policies, an Access Agreement, and an Access

Application Form [85]. Such documents correspond to the consent form and will require routine updates. Population biobanks include both individual and aggregate data in their collection. The latter can be made available online for researchers in an open access system. The former will require the creation of a controlled system, one in which applicants are required to submit an access application and have their request evaluated by pertinent access bodies [86,87]. Generally, a data and sample access committee will be constituted to adjudicate access requests and control the sharing of sensitive data [83]. These committees are typically composed of experts with backgrounds in epidemiology, law, ethics, and Information Technology [88]. If a project spans across multiple jurisdictions, a consolidation of access requests toward an access office will help to streamline the process [89,90].

In order to protect the privacy of participants, transferred data and samples will be coded. In addition, access agreements signed by approved researchers (and their institutions) will list a number of conditions that include, but are not limited to, prohibitions on both reidentifying participants and sharing data with unauthorized parties [91]. This agreement will also include a clause requiring the return of enriched data by approved users to the biobank [91]. Enriched data are data that are produced by the approved user as part of their project. Their return to the biobank will allow the population study to enhance its collection and offer future researchers a richer selection of variables for study [92]. Finally, given that broad consent is used in numerous population biobanks, some form of ongoing communication with participants will be undertaken [93]. When it comes to access mechanisms, this may be realized in the form of a public registry that can include researcher information and lay summaries of projects currently using resources provided by the population biobank [90,94]. This will allow participants to remain generally informed of how their data and samples are being used. In some cases, this may prompt them to withdraw participation. In fact, it is recommended that access bodies responsible for the adjudication of access requests identify what they consider as potentially objectionable research uses prior to allowing access to data and samples [95]. Not only would this require understanding the perceptions of research participants, but of the general public as well. This, once again, highlights the important role played by society in population biobanks [95].

2.7 Conclusion

This chapter introduced the several Canadian population biobanks that will be referred to throughout the remainder of this book. Using document analysis, I described the central characteristics of population biobanks and how they differ from other research projects. Four characteristics were noted in particular: (1) population biobanks are created with the goal of mainly benefitting the public and future generations; (2) they are established to supply data and samples for future research projects; (3) they are linked with administrative health data; and, finally, (4) they consist of organized and searchable collections. From the analysis of internal biobank documents, it became clear that there is a critical role played by the public and research community in population biobanks. Indeed, these projects are essentially created for the benefit of society, facilitated by the collection of data and samples and their linkage to administrative health data. Moreover, the fact that population biobanks maintain organized and searchable collections of data and samples that are accessible by the general research community increases the tangible role and impact played by researchers who apply for access. Understanding the crucial role played by stakeholders outside of the participant—researcher paradigm will work to demonstrate how individualistic autonomy is limited in the context of population biobanks. In particular, it will show how this particular conception of autonomy is unable to account for the multilateral relationships implicated in population research projects, including those that involve the broader research community and the general public. Furthermore, understanding that the public and research communities play important and meaningful roles will assist in the reexamination of how information will be disclosed to research participants in the future. This understanding will, I argue, ground an alternative conception of autonomy that does not see participants as independent agents but as interdependent with other stakeholders in a complex web of relations.

After having explored the distinctive characteristics of population biobanks using documents that reflect what research participants are provided in terms of information, I will now analyze how policies, guidelines, and statements have addressed the duty to inform of researchers in population biobanks, with particular attention given to the ways they approach situations in which researchers are limited in the information to participants.

References

[1] Khoury MJ. The case for a global human genome epidemiology initiative. Nat Genet 2004;36(10):1027—8.

[2] Awadalla P, Boileau C, Payette Y, Idaghdour Y, Goulet JP, Knoppers B, et al. Cohort profile of the CARTaGENE study: Quebec's population-based biobank for public health and personalized genomics. Int J Epidemiol 2013;42(5):1285—99.

[3] Gibbons SM, Kaye J, Smart A, Heeney C, Parker M. Governing genetic databases: challenges facing research regulation and practice. JL Soc 2007;34(2):163—89.

[4] Collins FS, Morgan M, Patrinos A. The Human Genome Project: lessons from large-scale biology. Science 2003;300(5617):286—90.

[5] Knoppers BM, Zawati MH, Kirby ES. Sampling populations of humans across the world: ELSI issues. Annu Rev Genomics Hum Genet 2012;13:395—413.

[6] Sanderson SC. Genome sequencing for healthy individuals. Trends Genet 2013;29(10):556—8.

[7] Hawkins AK. Biobanks: importance, implications and opportunities for genetic counselors. J Genet Counsel 2010;19(5):423—9.

[8] Council of Europe, Committee of Ministers. Recommendation Rec (2006) 4 of the Committee of Ministers to member states on research on biological materials of human origin. Recommendation Adopted 15 March 2006 (958th meeting of the Ministers' Deputies), s.17,. 2006. https://wcd.coe.int/ViewDoc.jsp?id=977859. [Accessed 21 March 2016].

[9] CanPath Portal. BC generations project (British Colombia). 2015. https://portal. canpath.ca/mica/individual-study/bcgp (recruitment statistic updated as of January 2017). [Accessed 21 March 2011].

[10] BC Generations Project. The project. 2019. http://www.bcgenerationsproject.ca/. [Accessed 21 March 2011].

[11] CanPath Portal. Alberta's tomorrow project (Alberta). 2015. https://portal.canpath.ca/ mica/individual-study/atp (recruitment statistic updated as of February 2015) [Accessed 21 March 2011].

[12] Count me in 4 tomorrow. Brief history & summary. 2012. http://in4tomorrow.ca/. [Accessed 21 March 2011].

[13] CARTaGENE. About. 2016. https://cartagene.qc.ca/en/about. [Accessed 21 March 2011].

[14] CanPath Portal. Ontario health study (Ontario). 2015. https://portal.canpath.ca/mica/ individual-study/ohs (recruitment statistic updated as of March 2018) [Accessed 21 March 2011].

[15] Ontario Health Study. About the study. 2018. https://www.ontariohealthstudy.ca/. [Accessed 21 March 2011].

[16] CanPath Portal. Atlantic PATH (Atlantic Region). 2018. https://portal.canpath.ca/mica/ individual-study/atlantic-path (recruitment statistic updated as of March 2018) [Accessed 21 March 2011].

[17] Atlantic PATH. Our study. 2018. http://atlanticpath.ca/. [Accessed 21 March 2011].

[18] Canadian alliance for healthy hearts & minds. Timetable: Release of Data; 2017. http:// cahhm.mcmaster.ca/?page_id=4278. [Accessed 21 March 2011].

[19] Canadian alliance for healthy hearts & minds. Research. 2019. https://www.phri.ca/ research/cahhm/. [Accessed 21 March 2011].

[20] Canadian Institutes of Health Research. Canadian longitudinal study on aging (CLSA). Government of Canada; 2018. http://www.cihr-irsc.gc.ca/e/18542.html (data collection infrastructures include: National Coordinating Centre (Hamilton, ON), Biorepository and Bioanalysis Centre (Hamilton, ON), Statistical Analysis Centre (Montreal, QC), Genetics and Epigenetics Centre (Vancouver, BC), 11 Data Collection Sites (Victoria, BC; Vancouver, BC; Surrey, BC; Calgary, AB; Winnipeg, MB; Hamilton ON; Ottawa ON; Montreal QC; Sherbrooke, QC; Halifax, NS; and St. John's, NL), 4 Computer-Assisted Telephone Interview Centres (Victoria, BC; Winnipeg MB; Sherbrooke, QC; and Halifax, NS), Information Technology Hub (Hamilton, ON)) [Accessed 21 March 2011].

[21] Canadian longitudinal study on aging. About the study, online. 2018. https://www.clsa-elcv.ca/. [Accessed 21 March 2011].

[22] Canadian longitudinal study on aging. Governance. 2018. https://www.clsa-elcv.ca/ about-us/governance. [Accessed 21 March 2011].

[23] Parodi B. Biobanks: a definition. Ethics, law and governance of biobanking. Cham: Springer; 2015. p. 15—9.

[24] Slokenberga S. Setting the foundations: individual rights, public interest, scientific research and biobanking. GDPR and biobanking. Cham: Springer; 2020. p. 11—30.

[25] National Institutes of Health (NIH). Genetics home reference website: "what are complex multifactorial disorders?". 2018. http://ghr.nlm.nih.gov/handbook/mutationsand disorders/complexdisorders. [Accessed 21 March 2011].

[26] Swede H, Stone CL, Norwood AR. National population-based biobanks for genetic research. Genet Med 2007;9(3):141—9.

[27] Knoppers BM, Harris JR, Burton PR, Murtagh M, Cox D, Deschênes M, et al. From genomic databases to translation: a call to action. J Med Ethics 2011;37(8):515—6.

[28] UK Biobank. UK Biobank. 2018. www.ukbiobank.ac.uk. [Accessed 21 March 2009].

[29] The Estonian Genome Centre. University of Tartu. www.geenivaramu.ee/en/. [Accessed 09 March 2009].

[30] Healthy Ageing Campus Groningen. Healthy ageing campus. 2018. https://campus. groningen.nl/about-campus-groningen/healthy-ageing-campus. [Accessed 21 March 2009].

[31] Biobanking and Biomolecular Resources Research Infrastructure (BBMRI) BBMRI— ERIC. 2018. http://www.bbmri-eric.eu/. [Accessed 21 March 2009].

[32] Canadian Partnership for Tommorow's Health (CanPath). CanPath cohort profile. 2021. https://canpath.ca. [Accessed 21 March 2009].

[33] Burton PR, Hansell AL, Fortier I, Manolio TA, Khoury MJ, Little J, et al. Size matters: just how big is BIG? Quantifying realistic sample size requirements for human genome epidemiology. Int J Epidemiol 2009;38(1):263—73.

[34] Ouellette S, Tassé AM. P3G—10 years of toolbuilding: from the population biobank to the clinic. Appl Trans Genomics 2014;3(2):36—40.

[35] Thorogood A, Joly Y, Knoppers BM, Nilsson T, Metrakos P, Lazaris A, et al. An implementation framework for the feedback of individual research results and incidental findings in research. BMC Med Ethics 2014;15(1):1—13.

[36] McGregor TL, Van Driest SL, Brothers KB, Bowton EA, Muglia LJ, Roden DM. Inclusion of pediatric samples in an opt-out biorepository linking DNA to De-identified medical records: pediatric BioVU. Clin Pharmacol Ther 2013;93(2):204—11.

[37] PopGen International Database. Population Biobanks Lexicon, a collaborative endeavour between: Public Population Project in Genomics and Society (P3G) & Promoting Harmonization of Epidemiological Biobanks in Europe (PHOEBE), Glossary: biobank. http://www.popgen.info/glossary. [Accessed 09 March 2021].

[38] Fransson MN, Rial-Sebbag E, Brochhausen M, Litton J-E. Toward a common language for biobanking. Eur J Hum Genet 2015;23(1):22—8.

[39] Shaw D, Elger BS, Colledge F. What is a biobank? Differing definitions among biobank stakeholders. Clin Genet 2014;85(3):223—7.

[40] Hansson MG. Striking a balance between personalised genetics and privacy protection from the perspective of GDPR. In: Slokenberga S, Tzortzatou O, Reichel J, editors. GDPR and biobanking: individual rights, public interest and research regulation across Europe. Cham: Springer; 2021. p. 45—60.

[41] Skipper M. The peopling of Britain. Nat Rev Genet 2015;16(5):256—7.

[42] Knoppers BM, Zawati MH. Population biobanks and access. In: Canestrari S, Zatti P, editors. Il governo del corpo: Trattato di biodiritto. Giuffrè Editore, vol. 2. Milan: Giuffrè Editore; 2011. p. 1181.

[43] Bowen GA. Document analysis as a qualitative research method. Qual Res J 2009: 27—8.

[44] Cancer Care Manitoba. CCMB tomorrow project. 2018. http://www.cancercare.mb.ca/ resource/File/CCMB-Tmrw-Proj_pamphlet_FNL_R1_web.pdf. [Accessed 21 March 2011].

[45] BC Generations Project. Consent form. British Columbia; 2014. p. 3—5 (obtained through correspondence).

[46] The tomorrow project. Alberta: Consent Form; 2011. p. 3 (obtained through correspondence).

[47] The tomorrow project. Alberta: Study Booklet; 2011. p. 4—6 (obtained through correspondence).

[48] Ontario Health Study. Consent form (obtained through correspondence). 2014.

[49] Ontario Health Study. Website FAQ. 2014. https://www.ontariohealthstudy.ca/about-the-study/frequently-asked-questions/. [Accessed 21 March 2011].

[50] CARTaGENE. Second wave information brochure for participants. 2014. https://cartagene.qc.ca/sites/default/files/documents/consent/cag_2e_vague_brochure_en_v3_7apr2014.pdf. [Accessed 21 March 2011].

[51] Atlantic PATH. Consent and brochure (obtained through correspondence). 2013. p. 2—4.

[52] Canadian Longitudinal Study on Aging. Study information package — home interview & data collection site visit. p. 2-9. https://www.clsa-elcv.ca/doc/414. [Accessed 21 March 2017].

[53] Canadian longitudinal study on aging, consent form — home interview & data collection site visit. https://www.clsa-elcv.ca/doc/448. [Accessed 21 March 2017].

[54] Canadian Alliance for Healthy Hearts and Minds. Participant information and consent Sheet (Aboriginal participants). (obtained through correspondence).

[55] Guion LA, Diehl DC, McDonald D. Triangulation: establishing the validity of qualitative studies. Environ Data Inf Serv 2011;2011(8):3.

[56] Master Z, Nelson E, Murdoch B, Caulfield T. Biobanks, consent and claims of consensus. Nat Methods 2012;9(9):885—8.

[57] Steinsbekk KS, Myskja BK, Solberg B. Broad consent versus dynamic consent in biobank research: is passive participation an ethical problem? Eur J Hum Genet 2013;21(9):897—902.

[58] Fonds de la recherche en santé du Québec. 2006. p. 55—6. Final Report — advisory Group on a Governance Framework for Data Banks and Biobanks Used for Health Research, www.frsq.gouv.qc.ca/en/ethique/pdfs_ethique/Rapport_groupe_conseil_anglais.pdf. [Accessed 10 March 2021].

[59] Tomlinson T, De Vries R, Ryan K, Kim HM, Lehpamer N, Kim SY. Moral concerns and the willingness to donate to a research biobank. J Am Med Assoc 2015;313(4):417—8.

[60] Canadian Institutes of Health Research. Natural Sciences and Engineering Research Council of Canada & social Sciences and Humanities Research Council of Canada, Tri-Council Policy Statement: Ethical Conduct for Research Involving Humans (Ottawa: Secretariat Responsible for the Conduct of Research, 2014) Art. 12.

[61] See e.g. BC Generations, Alberta Tomorrow Project, Ontario Health Study, CARTaGENE, AtlanticPATH in Table 2.

[62] Knoppers BM, Dove ES, Litton J-E, Nietfeld J. Questioning the limits of genomic privacy. Am J Hum Genet 2012;91(3):577.

[63] Ewing AT, Erby LA, Bollinger J, Tetteyfio E, Ricks-Santi LJ, Kaufman D. Demographic differences in willingness to provide broad and narrow consent for biobank research. Biopreserv Biobanking 2015;13(2):98.

[64] Council of Canadian Academies. Accessing health and health-Related data in Canada: executive summary. Ottawa: Council of Canadian Academies; 2015. http://www.scienceadvice.ca/uploads/eng/assessments%20and%20publications%20and%20news%20releases/Health-data/HealthDataExecSumEn.pdf. [Accessed 21 March 2011].

[65] Doiron D, Raina P, Fortier I. Linking Canadian population health data: maximizing the potential of cohort and administrative data. Can J Public Health 2013;104(3):e258—61.

[66] Wellcome Trust. Enabling data linkage to maximise the value of public health researh data: full report. 2015. p. 6. https://cms.wellcome.org/sites/default/files/enabling-data-linkage-to-maximise-value-of-public-health-research-data-phrdf-mar15.pdf. [Accessed 21 March 2017].

[67] Sudlow C, Gallacher J, Allen N, Beral V, Burton P, Danesh J, et al. UK biobank: an open access resource for identifying the causes of a wide range of complex diseases of middle and old age. PLoS Med 2015;12(3):e1001779.

[68]   Canadian Alliance for Healthy Hearts and Minds. participant information and consent sheet (MHI site) (obtained through correspondence).

[69]   McGill University — Faculty of Medicine. General guidelines for biobanks and associated databases. 2015. p. 4. https://www.mcgill.ca/medresearch/files/medresearch/guidelines_for_biobanks_and_associated_databases.march2015.pdf. [Accessed 21 March 2011].

[70]   Canadian Partnership for Tomorrow. Project's access policy (Approved May 11, 2016), Section 17. https://portal.partnershipfortomorrow.ca/sites/portal-live-7.x-5.10-020320 171455–partnershipfortomorrow.ca/files/Access_Policy_Approved_May_11_final.pdf>.

[71]   Shabani M, Borry P. You want the right amount of oversight: interviews with data access committee members and experts on genomic data access. Genet Med 2016;18(9): 892—7.

[72]   Deschenes M, Sallée C. Accountability in population biobanking: comparative approaches. J Law Med Ethics 2005;33(1):40—53.

[73]   Canadian Tumour Repository Network (CTRNet). Standard operating procedures. https://www.ctrnet.ca/en/resources/national-standards/. [Accessed 21 March 2011].

[74]   Bédard K, Wallace S, Lazor S, Knoppers B. Potential conflicts in governance mechanisms used in population biobanks. In: Kaye J, Stranger M, editors. Principles and practice in biobank governance; 2009. p. 217—27.

[75]   Canadian Institutes of Health Research. Advisory committee on ethical, legal and social issues for the CLSA. 2016. http://www.cihr-irsc.gc.ca/e/40803.html. [Accessed 21 March 2011].

[76]   Canadian Partnership for Tomorrow Project (CPTP). Canadian Partnership for Tomorrow project. 2018. www.partnershipfortomorrow.ca. [Accessed 21 March 2011].

[77]   Godard B. Involving communities: a matter of trust and communication. Cross Over-Genomics Public Arena 2005;93.

[78]   O'Doherty KC, Hawkins AK, Burgess MM. Involving citizens in the ethics of biobank research: informing institutional policy through structured public deliberation. Soc Sci Med 2012;75(9):1605.

[79]   Burgess M, O'Doherty K, Secko D. Biobanking in British Columbia: discussions of the future of personalized medicine through deliberative public engagement. 2008. p. 285.

[80]   UK Biobank. Newsletter 2015; 2015. https://web.archive.org/web/20200929155649/http://www.ukbiobank.ac.uk/newsletter-2015/. (archived website).

[81]   Knoppers B, Abdul-Rahman MZ. Biobanks in the literature. In: Bernice E, editor. Ethical issues in governing biobanks: global perspectives. Farnham: Ashgate Publishing; 2008. p. 13.

[82]   Prainsack B, Buyx A. A solidarity-based approach to the governance of research biobanks. Med Law Rev 2013;21(1):85.

[83]   Shabani M, Knoppers BM, Borry P. From the principles of genomic data sharing to the practices of data access committees. EMBO Mol Med 2015;7(5):507—9.

[84]   Lemmens T, Austin LM. The end of individual control over health information: promoting fair information practices and the governance of biobank research. In: Governing biobanks. Farnham: Ashgate; 2009. p. 250—1.

[85]   CanPath's Access Portal Documents. Data Access Policy, a Publications Policy, an Intellectual Property Policy and a Data Access Application Form. https://portal.canpath.ca/user/login?destination=node/7. [Accessed 21 March 2011].

[86]   Kaye J. Biobanking networks-what are the governance challenges? In: Kaye J, Stranger M, editors. Principles and practice in biobank governance. Farnham. UK: Ashgate; 2009. p. 201.

[87] Hallinan D. Biobank oversight and sanctions under the general data protection regulation. GDPR and biobanking. Cham: Springer; 2021. p. 123–7.

[88] Shabani M, Knoppers B, Borry P. Genomic databases, access review, and data access committees. In: Kumar D, Antonarakis S, editors. Medical and health genomics. Amsterdam: Elsevier; 2016. p. 32.

[89] Joly Y, Dove ES, Knoppers BM, Bobrow M, Chalmers D. Data sharing in the post-genomic world: the experience of the international cancer genome consortium (ICGC) data access compliance office (DACO). PLoS Comput Biol 2012;8(7): e1002549.

[90] Shabani M, Dyke SO, Joly Y, Borry P. Controlled access under review: improving the governance of genomic data access. PLoS Biol 2015;13(12):e1002339.

[91] Knoppers BM, Chisholm RL, Kaye J, Cox D, Thorogood A, Burton P, et al. A P3G generic access agreement for population genomic studies. Nat Biotechnol 2013; 31(5):384–5.

[92] Harris JR, Haugan A, Budin-Ljøsne I. Biobanking: from vision to reality. Norsk epidemiologi 2012;21(2).

[93] Platt J, Bollinger J, Dvoskin R, Kardia SL, Kaufman D. Public preferences regarding informed consent models for participation in population-based genomic research. Genet Med 2014;16(1):11–8.

[94] The International Cancer Genome Consortium's Data Access Application Office—DACO. DACO approved projects. 2021. https://daco.icgc.org/controlled-data-users/#/?&order=asc&by=proj. [Accessed 21 March 2010].

[95] Shabani M, Dove ES, Murtagh M, Knoppers BM, Borry P. Oversight of genomic data sharing: what roles for ethics and data access committees? Biopreserv Biobanking 2017;15(5):469–71.

The duty to inform of researchers in population biobanks

3.1 Introduction

In Chapter 1, I demonstrated that an individualistic conception of autonomy is at the core of the jurisprudential interpretation of the duty to inform in Canada. One consequence of this is that the duty to inform in research is more exacting than in the clinic setting. Participants, accordingly, have a right to receive "full and frank disclosure of all the facts, opinions, and probabilities" [1] during their consent to a research project. This standard is binding on all researchers working with human participants in Canada. At present, there is no legislation or case law that specifically provides an alternative standard for population biobanks. For this reason, population biobank researchers are expected to abide by the same exacting standard followed by researchers in other contexts.

This chapter will address the gap created by the absence of specific Canadian legislative guidance on population biobanking. It will do so by examining the range of internationally adopted guidelines, statements, policies, and legislation that address the provision of information to biobank participants. This comparative analysis will provide an account of what biobank researchers are expected to disclose to participants in the international setting. From this, in turn, I will draw comparisons between such expectations and the exacting standard demanded by Canadian courts. I conclude by outlining the various practical limitations faced by population biobank researchers when providing information to research participants during the consent process. In doing so, I draw upon the consent documents reviewed in Chapter 2. Understanding such limitations will be critical in my later work of assessing the feasibility of the individualistic conception of autonomy supported by Canadian courts and its application to population biobanks.

3.2 Methodology

The documents I review in this chapter were collected using the PopGen module, a comprehensive international database of legislation and policies relevant in population genetics [2]. The database contains more than 1000 documents, including

Reciprocity in Population Biobanks. https://doi.org/10.1016/B978-0-323-91286-0.00004-6

policies, statements, legislation, and regulations. They are categorized into three main groups. The first category is composed of documents that have been adopted by international bodies such as the World Medical Association (WMA) and UNESCO. The second category of documents are regional. These are policies, statements, and regulations adopted by institutions of the European Union, such as the European Parliament. Finally; the third category consists of national documents emanating from more than 100 countries on five continents (Europe, Asia, Africa, North, and South America). These are documents adopted by a legislative body or organization within a country and are applicable only within that jurisdiction.

I conducted an initial search for documents using the PopGen search engine. I searched for documents enacted between 1990 and 2017, with 1990 being PopGen's default set range. For both international and regional documents, I selected the keywords "research" and "consent," to ensure that I would get results that pertain to population biobanks or to research in general. I used the same keywords at the national level and obtained a very large number of documents (more than 300), which is distinct from results at the international and regional levels, for which relatively fewer documents were found. In order to control for documents applicable in the population biobanking context, I narrowed the search by adding the specific keyword: "biobank." Following my initial search, the PopGen search engine was discontinued. To account for significant developments in the field, I conducted an updated search using publicly available online resources in 2021, using the search terms outlined above. I further cross-referenced included documents against the United States Department of Health and Human Services (HHS) *International Compilation of Human Research Standards*, a list of over 1000 laws, regulations, and guidelines enacted internationally and in countries around the world [3]. The HHS compilation includes general research standards, genetics research standards, and governance documents for human biological materials. No additional documents were added to my review as a result of this search.

Overall, my review returned 22 documents. Of these, I selected only those that included guidance on the provision of information by researchers and that were either (1) seminal to (i.e., having international or regional outreach), though not specifically mentioning, biobanks or (2) specifically applicable in the population biobanking context. As a result of that triage, I excluded a total of 5 documents. The remaining 17 documents were then thoroughly assessed and instances of guidance on consent and information provision were identified using document analysis. More specifically, I evaluated these documents for indications of the types of applicable consent procedures and for any guidance on the kinds of information that participants should be provided. In the following three sections, I describe the results of the comparative analysis.

3.3 International documents[1]

A comparative review of international normative documents reveals that there has been consistent discussion of the responsibility of researchers to provide adequate information since the second half of the 20th century. Emerging in the *Nuremberg Code* [4] of 1949, "the duty and responsibility for ascertaining the quality of the consent rests upon each individual who initiates, directs, or engages in the experiment" [4]. The *Code* affirmed that this is a personal duty, "which may not be delegated to another with impunity" [4]. In this early iteration of the duty to inform, a clear link is drawn between the duty and the quality of participant consent. Quality here refers to the quality of the information provided during consent. A similar position is taken in the 2013 *Declaration of Helsinki* [5], which stipulates that

> *each potential subject must be adequately informed of the aims, methods, sources of funding, any possible conflicts of interest, institutional affiliations of the researcher, the anticipated benefits and potential risks of the study and the discomfort it may entail, post-study provisions, and any other relevant aspects of the study [5].*

In this *Declaration*, the provision of information is not only expected to occur during the initial consent phase, but in later phases of research as well. Indeed, article 26 additionally requires that research participants be given an opportunity to express their preferences about receiving further information about the general outcome and results of the study. Use of words such as "general" and "outcome" predicts the possibility of disclosure at the conclusion of the research project. UNESCO's *International Declaration on Human Genetic Data* [6] includes a similarly structured duty to inform, which includes a right of participants to decide whether they wish to be informed of research results [6]. This position has been taken consistently by UNESCO since the 1997 adoption of the *Universal Declaration on the Human Genome and Human Rights* [6].

The Council for International Organizations of Medical Sciences [CIOMS] has adopted *International Ethical Guidelines for Biomedical Research Involving Human Subjects* [7]. This document states that researchers have a duty to "seek and obtain consent, but only after providing relevant information about the research and ascertaining that the potential participant has adequate understanding of the material facts" [8]. More importantly, however, it also acknowledges that seeking specific consent when future use remains uncertain will be challenging [9].

Similarly, the WMA's *Declaration of Taipei on Ethical Considerations Regarding Health Databases and Biobanks* [10] takes the position that researchers should, by default, always obtain the specific, free, and informed consent of

---

[1] Portions of this section have previously appeared in Zawati MH, There Will be Sharing: Population Biobanks, the Duty to Inform, and the Limitations of the Individualistic Conception of Autonomy. Health LJ. 2014; 21.

participants for the storage, collection, and use of data and samples [5]. According to the WMA, in cases of predicted indefinite use, consent may only be valid if participants are informed about a range of issues, including, but not limited to, the nature of the data or sample to be collected, how participant privacy will be protected, the nature of the governance arrangements of the biobank, the procedures for the return of results, and rules for accessing data and samples [10]. This approach was inspired by the OECD's 2009 *Guidelines on Human Biobanks and Genetic Research Databases* [11].

3.4 Regional documents[2]

Regional normative instruments are broadly similar to international documents in their treatment of information provision. The seminal *Convention on Human Rights and Biomedicine* [12] (*Oviedo Convention*) of the Council of Europe states that participants in a research project "shall beforehand be given appropriate information as to the purposes and nature of the intervention as well as on its consequences and risks" [13]. This principle is reiterated in various other European norms, such as *Directive 2001/20/EC* [14] and the *Recommendation Rec (2006) 4 of the Committee of Ministers to member states on research on biological materials of human origin* [15], which explicitly discusses population studies.

Article 10 of the *Oviedo Convention* recognizes a "right to information," such that participants "[are] entitled to know information collected about [their] health" unless they explicitly invoke their right not to be informed. Importantly, this right not to know is never absolute, and may be restricted in the interests of the participant in question [12]. This constraint may be applied, for example, where clinically significant information is discovered about a juvenile participant that may be actionable during childhood [16].

Likewise, the *Additional Protocol to the Convention on Human Rights and Biomedicine, concerning Biomedical Research* (*Additional Protocol*) [17], emphasizes the importance of providing participants with sufficient information in a comprehensible form. It confirms that patients are entitled to know any collected information that concerns their health [18]. To accomplish this, the *Additional Protocol* creates a "duty of care" on the part of researchers to communicate relevant information in the case that a study "gives rise to information of relevance to the current or future health or quality of life of research participants" [19]. The *Additional Protocol* specifies that such information must be disseminated through a framework of health care or counseling and that researchers are under an obligation to protect both the confidentiality of information and the wishes of participants [19].

---

[2] Portions of this section have previously appeared in Zawati MH, Liability and the Legal Duty to Inform. In: Joly Y, Knoppers BM. Routledge Handbook of Medical Law and Ethics. London: Routledge; 2014. p. 199−200.

3.5 **National documents**

At the beginning of the 21st century, Iceland became the first European country to adopt legislation specifically directed at biobanks. The 2000 *Act on Biobanks* [20] requires that biological samples collected for storage in a research biobank be accompanied by the free and informed consent of the donor [21]. It adds that "[t] his consent shall be given freely and in writing after the donor of a biological sample has been informed of the objective of the sample collection, the benefits, risks associated with its collection […] [21]." Estonia's 2000 *Human Genes Research Act* [22], in turn, states that it is "prohibited to take a tissue sample and prepare a description of state of health or genealogy without the specific knowledge and voluntary consent of the person" [23]. Sweden's 2002 *Biobanks in Medical Care Act* [24] focuses on the importance of informing participants about the intention and purpose of a biobank project [25]. The *Act* further insists that tissue samples stored in biobanks may not be used for purposes other than those indicated in consent documents [26]. Legislation enacted in Belgium [27], Finland [28], and Taiwan [29] ha all also included similar elements.

In Canada, the *Tri-Council Policy Statement* (TCPS 2)—a research ethics document that is binding on researchers funded by one of the three councils—includes a chapter dedicated entirely to human biological materials. That chapter sets out a number of requirements for consent [30]. Apart from referring to elements of consent set out in article 3.2 (including purpose, risks, benefits, and others), article 12.2 of the *Statement* demands that researchers disclose, among other things, "the manner in which biological materials will be taken, […] the safety and invasiveness of the procedures for acquisition," the intended use and plans for "handling results and findings, including clinically relevant information and incidental findings" [30]. These directives on the return of clinically relevant information are supplemented by the 2019 guidance document "How to Address Material Incidental Findings" [31]. Among other things, this guidance identifies the population biobank setting as one instance in which the circumstances may justify an exception to the general obligation to disclose material incidental findings, particularly when research participants consent to a policy that does not permit the return of results [32]. The TCPS 2 further acknowledges that some biological materials will be collected for research purposes and may also be used in "future research, although the precise research project(s) may not be known at the time" [33]. This statement seems to acknowledge practical limitations on the part of researchers, namely, that they are unable to foresee future use at the time of consent. Although interesting, this statement does not override the standard outlined in case law. Furthermore, the lack of elaboration by the TCPS 2 on this point indicates that its interpretation is unsettled.

In 2010, the German National Ethics Council adopted a guidance document entitled *Human Biobanks for Research* [34]. This document adopts a general position on consent that bears some resemblance to those reviewed above. The guidance states that consent must be preceded by appropriate information about the purpose, significance, and implications of the research project [34]. This, according to the document, presupposes specific consent. In a manner similar to various more recent international documents (such as the CIOMS and the WMA's *Declaration of Taipei*), the German National Ethics Council suggests that, where specific consent is impossible, consent documents must include sufficient information related to the kinds of materials and data to be collected, how such collections will be stored, to whom materials and data will be provided, and how the collection will be protected [35]. In the same vein, the United States 2017 revised *Common Rule* [36], a national research ethics document, also recognized this broader form of consent for biobanks, albeit with some conditions [36].

By examining the various international, regional, and national normative documents reviewed in the section above, several conclusions can be drawn. First, most international and regional documents do not consider population biobanks specifically. Rather, they take general positions on the importance of providing adequate information to research participants. More recent international documents have included greater elaboration on the duty of researchers to inform in biobanking research. National documents usually take the same approach, with greater emphasis placed on issues associated with biobanking in particular, including the importance of providing participants with information on the future use and storage of data and samples. Second, and perhaps more importantly, the comparative analysis of documents has shown that a number of jurisdictions have provided guidance on how much information should be provided to participants in the research biobanking context. Cognizant of limitations faced by biobanks in providing participants with specific information, some of these documents have instead focused on protecting the confidentiality of data and samples. However, the same cannot be said about the approach taken in Canadian law. As I demonstrated above, Canadian law requires that research participants are informed of all facts, opinions, and probabilities prior to giving research consent. This, I argue, inevitably places unreasonable limitations on researchers in the biobanking context. Many such limitations are likely impossible to satisfy. In the following section, I will illustrate them.

3.6 Limitations to the duty to inform in the context of population biobanks

The range of norms examined above reflect a trend in guidance on consent and the duty to inform: researchers must adequately inform participants about the risks, goals, and potential outcomes of research projects during the consent process. The precise nature and content of the required consent, however, remain unsettled in the field of population biobanking [37—40]. While we have seen that some international and national documents propose solutions when researchers are unable to provide specific information, the same cannot be said of Canadian Courts. This is why it is critical to more precisely understand the practical limitations faced by population biobanks when disclosing information to research participants.

Given that population biobanks are designed to foster future research, there are a certain number of inevitable limits on what may feasibly be disclosed to participants. In the population biobanking context, future users and specific proposed research projects are unknown at the time of initial consent [41]. Biobank researchers will often find themselves unable to fully inform participants about the "intended uses" or the "range and duration" of such use at the moment of initial consent. On the other hand, requesting specific consent from participants where exacting information will be provided—such as information about the researchers who will have access to data and samples and the nature of their specific research project—will likely restrict future access to such data and samples. The reason for this is, simply put, that a process of reconsent would be required to follow every new access application. The process of reconsenting research participants in this way would be both costly and time consuming, owing largely to the high number of participants and the limited resources available to undertake such a reconsenting process [42]. Moreover, a process that includes reconsent may negatively impact recruitment efforts. Indeed, depending on the frequency of requests, there is a possibility that participants, exasperated from constant reconsenting efforts, will drop out of the biobank altogether [42]. This, in turn, would affect the long-term sustainability of the population biobank.

Given this situation, a number of population biobanks have resorted to the adoption of broad consent [43]. The term "broad consent" or "general consent" means "consenting to a framework for future research of certain types" [44] and pertains "to a bank or research infrastructure whose possible uses are not all known at the start" [45]. This category of consent is alluded to in some of the documents reviewed in this chapter, namely by the WMA, OECD and CIOMS, TCPS 2, German National Ethics Council, and the American 2017 revised *Common Rule*. Recent commenters have suggested that the enactment of the European General Data Protection Regulation (GDPR) similarly recognizes the necessity of broad consent for biobanking [46]. The GDPR permits the use of broad consent so long as it is accompanied by sufficient ethical oversight [47]. Some proponents of broad consent point to practical limitations listed above and support arguments in favor of this alternative approach by claiming that biobank participation is a relatively low-risk form of research participation [48]. This view, however, has not received unanimous agreement in the literature [45]. In fact, opponents of broad consent argue that one of the key elements of consent, namely, that it is *informed*, goes unsatisfied in broad consent regimes [46]. Scholars of that view claim that donors only receive "information on general categories of foreseeable problems […] and benefits […], but they get no information about the specific research that will be done with their samples […]" [49]. Other authors have gone as far as saying that biobanks, as a matter of fact, are not low-risk research enterprises, especially considering that there is a real possibility of reidentification of participants by third parties [50]. Adding to this problem, there is no consensus on the perspectives of members of the public or participants regarding the type of consent researchers in biobanking research should seek. Indeed, while some authors have shown that a majority of participants prefer one-time broad consent [51], others have demonstrated that it is either a close split decision or that there is a preference for specific consent [52,53]. These findings are insufficient to the extent that they lack nuance with respect to their consideration of issues related to biobanking. As I described above, a number of the surveyed documents make generalizations about biobanking in discussions of particular issues. Many do not, for example, specifically focus on one species of biobank, such as population biobanks. This presents the risk that critical characteristics of the biobank under study will not be captured and that problems particular to a specific kind of biobanking project will be ignored. As a matter of fact, it is rare to encounter explicit discussions of broad and specific consent practices that contemplate the full diversity of biobanking projects. This is problematic to the extent that solutions proposed in one context are often inapplicable in others. A more fulsome discussion of this shortcoming, however, falls outside of the objectives of this chapter.

In an effort to defend the use of broad consent as a model for population biobanks, some proponents have maintained that as "long as […] broad consent is thorough and includes a discussion of the goals and relevant process" [54], such as the manner in which tissues will be conserved, mechanisms for ensuring the security of data, and ongoing governance structures for access and ethics monitoring [55], it could meet the broad requirements of informed consent [56]. Additionally, while the broad consent approach privileges flexibility, owing to its ability to envision a wider set of uses for data and samples, its promoters insist that such flexibility does not constitute a "carte blanche" [57]. Indeed, defenders argue that broad consent should be accompanied by additional security and governance mechanisms [55]. Beyond that, population studies that apply broad consent procedures often periodically recontact donors to administer questionnaires and collect additional samples, "thereby providing an opportunity for renewing consent and the right to withdraw through participant response over time" [58]. During such recontact procedures, consent forms that include any updated information are presented to participants. Participants are then given an opportunity to reassert whether they are interested in continuing their participation. Some authors have argued that iterative processes of this kind are indicative of a move toward a more dynamic consent model [42,59], one in which participants are provided "active opt-in requirements for each downstream research project" [44]. More precisely, dynamic consent is an online approach that may be put in place to accommodate different consent models depending on the objectives and context of the research project. In the future, participants can also benefit from this online system to consent to novel research studies or to modify initial consent along the way, thereby allowing for dynamic interactions between the participant and the researcher [59]. While I will not discuss this model in detail, I should mention that such dynamic consent has also received a fair share of critique in recent years [44].

As for population biobanks in the Canadian context, a review of consent forms and associated documents from such studies reveals that the broad consent approach described above is gradually being implemented. For all existing Canadian biobanks, the limited disclosure of unknown future access of data and samples is paired with rigorous governance and heightened privacy protection. As an example, the

Alberta Tomorrow Project discloses that participant data and samples "may be used, in coded form, by approved researchers from Canada and other countries for research related to cancer, and potentially other health conditions" [60]. Such potential use is based on the condition that prospective researchers apply for access under a controlled-access governance system [61]. CARTaGENE takes a similar approach, stating explicitly that "data and samples collected for the CARTaGENE project will be used for research on health and/or genomics" [62]. This kind of use is paired with the promise that an "ethics committees will evaluate the research projects submitted and the scientific validity of these studies will be examined by an access committee independent from CARTaGENE" [62].

Even granting that broad consent is, as its proponents suggest, a form of compromise between competing values; it remains unclear whether it is capable of being reconciled with legal requirements surrounding the duty to inform set by Canadian courts. Put another way, it is not evident that broad consent would satisfy the strict requirement to provide participants with a full and frank disclosure of all facts, opinions, and probabilities that is described in the *Halushka* and *Weiss* decisions. Recently, commentators have described the continued consent problem facing biobanks [63]. In a 2017 article, authors Caulfield and Murdoch state that "there remains a great deal of uncertainty regarding [...] what type of consent is legally appropriate" [63]. They conclude that broad consent does not appear to fulfill legal requirements in Canada and that "the time is now for policymakers and politicians to clear up the confusion" [63]. While I agree a problem exists and that it is time to dissipate confusion, I do not share in the conclusion that the issue applies to biobank consent, per se. Instead, I argue that the central concern turns on the individualistic conception of autonomy promoted by Canadian courts, which is the basis of the exacting duty imposed on biobank researchers. This is a claim that I will defend in the following Chapter.

3.7 Conclusion

This chapter had two objectives. First, it focused on the gap in specific legislative guidance related to population biobanks and examined the range of guidelines, statements, policies, and legislation that have been adopted internationally to address requirements surrounding the provision of information to biobank participants. To assuage the lack of specific Canadian guidance on this matter, I presented the results of an international comparative review of guidelines, statements, policies, and legislation that have been adopted on the topic of population biobanks. This review demonstrated that the requirement that sufficient and adequate information be provided to participants in biobanking research is widespread. More importantly, several of the documents analyzed have clearly recognized the limitations of specific consent and suggest a broader information provision requirement on the part of researchers. This stands in contrast to legal requirements in Canada demanding more directed, specific consent in which all opinions, probabilities, and facts are presented to the research participant.

Second, this chapter outlined various practical limitations faced by population biobank researchers when providing information to research participants during the consent process. Drawing on the consent forms and associated documentation reviewed in Chapter 2, I described the inability of population biobanks to foresee all possible uses of data and samples and the infeasibility of reconsenting participants every time a new project requests access to their data and samples. As a matter of course, this would require that population biobanks deviate from full disclosure requirements in Canadian law. From there, I briefly presented some of the potential solutions that have been discussed in the literature. Despite extensive discussion on the topic of biobanking and informed consent, there is some continued controversy on the best approach to follow when providing information to participants. This is so, I argue, primarily because many of the proposed solutions, such as broad consent or dynamic consent, are practical solutions generated by biobanks themselves, with limited conceptual support. The practical limitations of the individualistic conception of autonomy, however, require redress from a more theoretical point of view. In the following Chapter, I will argue that the shortcomings of individual autonomy are broader than the practical concerns identified here and, in fact, touch on more complex matters related to the multilateral nature of the research relationship in the context of population biobanks.

References

[1] Halushka v. University of Saskatchewan. 1965. 53 DLR (2d) 436 at 443−444, 52 WWR (ns) 608 (Sask CA).

[2] PopGen Module. International database on the legal and socio-ethical aspects of population genomics. 2017 [archived website], https://web.archive.org/web/20160315170944/www.popgen.info/home.

[3] Office for Human Research Protections. U.S. department of health and human Services, international compilation of human research standards. Washington: Department of Health and Human Services; 2020.

[4] Nuremberg Military Tribunals. Permissible medical experiments. In: Trials of war criminals before the Nuremberg military tribunals under control council law, vol 10:2. Washington: US Government Printing Office; 1949.

[5] World Medical Association. Declaration of Helsinki - ethical principles for medical research involving human subjects, 64th WMA general assembly. Fortaleza; 2013. https://www.wma.net/policies-post/wma-declaration-of-helsinki-ethical-principles-for-medical-research-involving-human-subjects/ [Accessed 21.03.11].

[6] UNESCOR. International declaration on human genetic data, 32nd sess, resolutions, item 22, SHS/BIO/04/1 REV. 2003.

[7] Council for International Organizations of Medical Sciences (CIOMS). International ethical guidelines for biomedical research involving human subjects. Geneva: WHO Press; 2016. https://cioms.ch/wp-content/uploads/2017/01/WEB-CIOMS-EthicalGuidelines.pdf.

[8] Council for International Organizations of Medical Sciences (CIOMS). International ethical guidelines for biomedical research involving human subjects, guideline 9. Geneva: WHO Press; 2016. https://cioms.ch/wp-content/uploads/2017/01/WEB-CIOMS-EthicalGuidelines.pdf.

[9] Council for International Organizations of Medical Sciences (CIOMS). International ethical guidelines for biomedical research involving human subjects, guideline 11. Geneva: WHO Press; 2016. https://cioms.ch/wp-content/uploads/2017/01/WEB-CIOMS-EthicalGuidelines.pdf.

[10] World Medical Association. WMA declaration of Taipei on ethical considerations regarding health databases and biobanks. 2016. https://www.wma.net/policies-post/wma-declaration-of-taipei-on-ethical-considerations-regarding-health-databases-and-bio banks/ [Accessed 21.03.11].

[11] OECD. Guidelines on human biobanks and genetic research databases, best practice 4.1. 2009. www.oecd.org/science/biotechnologypolicies/44054609.pdf. [Accessed 11 March 2021].

[12] Council of Europe. Convention for the protection of human rights and dignity of the human being with regard to the application of biology and medicine: convention on human rights and biomedicine. April 4, 1997. ETS No 164 (entered into force 1 December 1999).

[13] Council of Europe. Convention for the protection of human rights and dignity of the human being with regard to the application of biology and medicine: convention on human rights and biomedicine. April 4, 1997. ETS No 164, art 5 (entered into force 1 December 1999).

[14] EC. Directive 2001/20/Ec of the European parliament and of the council of 4 April 2001 [2001] OJ, L 212/34, art 3.

[15] Council of Europe, Committee of Ministers. Recommendation rec (2006) 4 of the committee of Ministers to member states on research on biological materials of human origin. Recommendation adopted 15 March 2006 (958th meeting of the Ministers' Deputies), s 17, art 14. 2006. https://wcd.coe.int/ViewDoc.jsp?id=977859 [Accessed 21.03.16].

[16] Hens K, Van El CE, Borry P, Cambon-Thomsen A, Cornel MC, Forzano F, et al. Developing a policy for paediatric biobanks: principles for good practice. Eur J Hum Genet 2013;21(1):2−6.

[17] Council of Europe. Additional Protocol to the convention on human rights and biomedicine, concerning biomedical research. January 25, 2005. ETS No195 (entered into force 1 October 2007), art 13(1).

[18] Council of Europe. Additional Protocol to the convention on human rights and biomedicine, concerning biomedical research. January 25, 2005. ETS No195 (entered into force 1 October 2007), art 26(1).

[19] Council of Europe. Additional Protocol to the convention on human rights and biomedicine, concerning biomedical research. January 25, 2005. ETS No195 (entered into force 1 October 2007), art 27.

[20] Act on Biobanks. 2000 (Iceland) no. 110. As amended by act no. 27/2008 and act no. 48/2009. Available from: https://www.government.is/media/velferdarraduneyti-media/media/acrobat-enskar_sidur/Biobanks-Act-as-amended-2015.pdf. [Accessed 21 03 12].

[21] Act on Biobanks. 2000 (Iceland) no. 110, as amended by act no. 27/2008 and act no. 48/2009, art 7. Available from: https://www.government.is/media/velferdarraduneyti-

media/media/acrobat-enskar_sidur/Biobanks-Act-as-amended-2015.pdf. [Accessed 21 03 12].

[22] Human genes research act 2000 (Estonia) RT I (104, 685). Available from: https://www. riigiteataja.ee/en/eli/531102013003/consolide. [Accessed 12 03 2021].

[23] Human genes research act 2000 (Estonia) RT I (104, 685), art 9. Available from: https:// www.riigiteataja.ee/en/eli/531102013003/consolide. [Accessed 12 03 2021].

[24] Biobanks in medical care act 2002 (Sweden). Available from: https://biobanksverige.se/ wp-content/uploads/Biobanks-in-medical-care-act-2002-297.pdf. [Accessed 21 03 11].

[25] Biobanks in medical care act 2002 (Sweden), Chapter 2 Sec. 5. Available from: https:// biobanksverige.se/wp-content/uploads/Biobanks-in-medical-care-act-2002-297.pdf. [Accessed 21 03 11].

[26] Biobanks in medical care act 2002 (Sweden), Chapter 3 Sec. 5. Available from: https:// biobanksverige.se/wp-content/uploads/Biobanks-in-medical-care-act-2002-297.pdf. [Accessed 21 03 11].

[27] Loi relative a l'obtention et a l'utilisation de matériel corporel humain destiné a des applications médicales humaines ou a des fins de recherche scientifique (Belgium) M.B. 30/12/2008. https://www.ieb-eib.org/fr/pdf/l-20081219-rech-mater-humain.pdf. [Accessed 21 03 11].

[28] Finnish biobank act 688/2012. Available from: http://www.finlex.fi/en/laki/kaannokset/ 2012/en20120688.pdf. [Accessed 21 03 15].

[29] Human biobanks management act 2012 (Republic of China). Hua-Zong-Yi-Yi-Tzu No 09900022481. http://law.moj.gov.tw/Eng/LawClass/LawAll.aspx?PCode=L0020164. [Accessed 21 03 15].

[30] Canadian Institutes of Health Research. Natural Sciences and Engineering Research Council of Canada & Social Sciences and Humanities Research Council of Canada. Tri-council policy statement: ethical conduct for research involving humans. Ottawa: Secretariat Responsible for the Conduct of Research; 2014. art. 12.

[31] Canadian Institutes of Health Research. Natural Sciences and Engineering Research Council of Canada & Social Sciences and Humanities Research Council of Canada. How to address material incidental findings: guidance in applying TCPS2 (2018). Ottawa: Secretariat on Responsible Conduct of Research; 2019. art 3.4.

[32] Canadian Institutes of Health Research. Natural Sciences and Engineering Research Council of Canada & Social Sciences and Humanities Research Council of Canada. How to address material incidental findings: guidance in applying TCPS2 (2018). Ottawa: Secretariat on Responsible Conduct of Research; 2019. art 4b(iii).

[33] Canadian Institutes of Health Research, Natural Sciences and Engineering Research Council of Canada & Social Sciences and Humanities Research Council of Canada. Tri-council policy statement: ethical conduct for research involving humans. Ottawa: Secretariat Responsible for the Conduct of Research; 2014. Chapter 12 B(3).

[34] Deuttscherr Etthiikrratt. Human biobanks for research. Berlin: Deuttscherr Etthiikrratt; 2010. p. 15.

[35] Deuttscherr Etthiikrratt. Human biobanks for research. Berlin: Deuttscherr Etthiikrratt; 2010. p. 37−8.

[36] United States Department of Human Health and Services. Final revisions to the common rule, federal register 82:12. 2017. https://www.gpo.gov/fdsys/pkg/FR-2017-01-19/pdf/2017-01058.pdf [Accessed 21.03.15].

[37] Allen J, McNamara B. Reconsidering the value of consent in biobank research. Bioethics 2011;25(3):155−66.

[38] Caplan AL. What no one knows cannot hurt you: the limits of informed consent in the emerging world of biobanking. In: Solbakk H, Holm S, Hofmann B, editors. The ethics of research biobanking. London: Springer; 2009.

[39] Beskow LM, Friedman JY, Hardy NC, Lin L, Weinfurt KP. Developing a simplified consent form for biobanking. PLoS One 2010;5(10):e13302.

[40] Caulfield T. Biobanks and blanket consent: the proper place of the public good and public perception rationales. King's LJ 2007;18(2):209–26.

[41] Genetics ESoH. Data storage and DNA banking for biomedical research: technical, social and ethical issues. Eur J Hum Genet 2003;11(12).

[42] Kaye J, Whitley EA, Lund D, Morrison M, Teare H, Melham K. Dynamic consent: a patient interface for twenty-first century research networks. Eur J Hum Genet 2015; 23(2):141–6.

[43] Master Z, Nelson E, Murdoch B, Caulfield T. Biobanks, consent and claims of consensus. Nat Methods 2012;9(9):885–8.

[44] Steinsbekk KS, Myskja BK, Solberg B. Broad consent versus dynamic consent in biobank research: is passive participation an ethical problem? Eur J Hum Genet 2013; 21(9):897–902.

[45] Fonds de la recherche en santé du Québec. Final report – advisory group on a Governance Framework for Data Banks and Biobanks Used for Health Research. 2006. p. 55–6. www.frsq.gouv.qc.ca/en/ethique/pdfs_ethique/Rapport_groupe_conseil_anglais.pdf. [Accessed 10 March 2021].

[46] Hansson MG. Striking a balance between personalised genetics and privacy protection from the perspective of GDPR. In: Slokenberga S, Tzortzatou O, Reichel J, editors. GDPR and biobanking: individual rights, public interest and research regulation across Europe. Cham: Springer; 2021. p. 45–60.

[47] Shabani M, Chassang G, Marelli L. The impact of the GDPR on the governance of biobank research in: GDPR and biobanking individual rights, public interest and research regulation across Europe. Cham: Springer; 2021. p. 31–44.

[48] D'Abramo F, Schildmann J, Vollmann J. Research participants' perceptions and views on consent for biobank research: a review of empirical data and ethical analysis. BMC Med Ethics 2015;16(1):1–11.

[49] Greely HT. The uneasy ethical and legal underpinnings of large-scale genomic biobanks. Annu Rev Genomics Hum Genet 2007;8:343–64.

[50] Stein DT, Terry SF. Reforming biobank consent policy: a necessary move away from broad consent toward dynamic consent. Genet Test Mol Biomarkers 2013;17(12): 855–6.

[51] Caulfield T, Rachul C, Nelson E. Biobanking, consent, and control: a survey of Albertans on key research ethics issues. Biopreserv Biobanking 2012;10(5):433–8.

[52] Ewing AT, Erby LA, Bollinger J, Tetteyfio E, Ricks-Santi LJ, Kaufman D. Demographic differences in willingness to provide broad and narrow consent for biobank research. Biopreserv Biobanking 2015;13(2):98.

[53] Nanibaa'A G, Sathe NA, Antommaria AHM, Holm IA, Sanderson SC, Smith ME, et al. A systematic literature review of individuals' perspectives on broad consent and data sharing in the United States. Genet Med 2016;18(7):663–71.

[54] Caulfield TA, Knoppers BM. Consent, privacy & research biobanks. Policy brief no. 1. Genomics, public policy & society 1 2010 (citing Ants Nõmper, open consent: a new form of informed consent for population genetic databases).

[55] Knoppers B, Abdul-Rahman MZ. Biobanks in the literature. In: Bernice E, editor. Ethical issues in governing biobanks: global perspectives. Farnham: Ashgate Publishing; 2008. p. 13.

[56] Knoppers BM, Abdul-Rahman MH. Health privacy in genetic research: populations and persons. Polit Life Sci 2009;28(2).

[57] Knoppers BM, Leroux T, Doucet H, Godard B, Laberge C, Stanton-Jean M, et al. Framing genomics, public health research and policy: points to consider. Public Health Genomics 2010;13(4):224−34.

[58] Caulfield TA, Knoppers BM. Consent, privacy & research biobanks. Genome Canada; 2010.

[59] Budin-Ljøsne I, Teare HJ, Kaye J, Beck S, Bentzen HB, Caenazzo L, et al. Dynamic consent: a potential solution to some of the challenges of modern biomedical research. BMC Med Ethics 2017;18(1):1−10.

[60] The tomorrow project. Alberta: Consent Form; 2011. p. 3 (obtained through correspondence).

[61] CPTP's Access Portal Documents. https://portal.partnershipfortomorrow.ca/request-access. [Accessed 10 March 2021].

[62] Atlantic PATH. Consent and brochure (obtained through correspondence). 2013. p. 2−4.

[63] Caulfield T, Murdoch B. Genes, cells, and biobanks: yes, there's still a consent problem. PLoS Biol 2017;15(7):2−6.

Limitations of the individualistic conception of autonomy in population biobanking

4.1 Limitations of the individualistic conception of autonomy: an introduction

In Chapter 1, I gave an overview of the evolution of the duty to inform in Canada. While paternalism was once a dominant norm in clinical practice, it was eventually replaced by respect for autonomy as the theoretical basis of the duty to inform in the second half of the 20th century. More importantly, I demonstrated that an individualistic conception of autonomy is at the core of the interpretation of the duty to inform by Canadian courts. These decisions have since informed our understanding of the duty to inform in nontherapeutic research, a duty that was determined to be more exacting than that of physicians in a clinical setting.

In Chapter 2, I examined the nature and characteristics of population biobanks and outlined how they differ from research projects considered in leading Canadian court decisions. Indeed, drawing on a review of internal documents presented to research participants by Canadian population biobanks, it can be seen that a much larger role is thought to be played by the public and—to some extent—the research community in this context. Not only do these projects recruit participants from the general population, but their governance is also established in the specific aim of maintaining public trust [1,2].

Chapter 3 described practical limitations that population biobank researchers face when providing information to participants. Among such limitations are the inability to foresee all possible uses of data and samples and the infeasibility of reconsenting participants each time a new project requests access. Both limitations would be actualized under a consent model motivated by individual autonomy. While neither the federal government nor any of the provinces have enacted legislation specifically regulating biobanks, Chapter 3 highlighted how guidelines, statements, and recommendations in other countries, in addition to those enacted by international and regional organizations, have recognized that specific models of consent are limited and have proposed a broader form of information provision by researchers.

For the time being, Canada's legal duty to inform continues to be based, at its core, on an individualistic conception of autonomy. But this conception faces several important theoretical shortcomings. Using information gathered in Chapters 2 and 3, this chapter will examine such shortcomings in detail.

Before doing so, it is important to note that individual autonomy has received a good deal of criticism by authors who have analyzed its inadequacies in the clinical setting. These inadequacies transcend the clinical setting to have important effects on research. One criticism is that autonomy is "highly individualistic" [3] in orientation. Several authors contend that this illustrates the manner in which "rights" may be claimed "without any sense of reciprocal obligations" [4]. Put another way, the relevant relationships are "unidirectional" in the sense that the role of a physician is limited to that of a passive provider of information [5], while little is said about possible patient obligations [4]. Others have gone so far as to qualify a patient–physician relationship based on individual autonomy as one of "bioethical paternalism," which leads "some doctors to consider mistakenly that unthinking acquiescence to a requested intervention against their clinical judgment is honoring 'patient autonomy' when it is, in fact, abrogation of their duty as doctors" [4]. In other words, "a competent patient's decision is good simply by virtue of having been made by the patient" [4]. Others still argue that traditional conceptions of autonomy focus on individuals as inherently independent, *atomistic* agents [6]. Conceiving of individuals as "bounded, ideally independent, and … regularly self-interested individuals," these authors suggest, is both founded on flawed assumptions about human behavior and has the effect of limiting progress in biomedical ethics [6]. All of these examples point to one important shortcoming: the individualistic conception of autonomy conceives of participants as fully independent individuals, entitled to information that would further their own interests. More to the point, accounts of individualistic autonomy typically fail to mention how the interactions of patients with others shape their decisions and, correlatively, how their decisions might affect others.

I argue that these shortcomings of individual autonomy transcend the clinical setting and have important implications for population biobanks. More concretely, I focus on two specific problems with individual autonomy in the population biobank setting. The first turns on how individual autonomy fails to recognize the complexities of benefit considerations in the research setting. The second, related to the first, considers how individual autonomy, with its unidirectional focus on the participant, is incapable of sustaining that same participant within the multilateral and complex relationships that involve the public and research community. Finally, this chapter will demonstrate that many of the proposed solutions to these shortcomings—namely deliberative autonomy, principled autonomy, and the duty to participate in research—do not resolve the limitations at issue. Relational autonomy, however, does represent a useful conception that could conceivably be adapted to the population biobank setting.

4.2 **The concept of benefit: moving beyond individual participants**

The research relationship described by Canadian courts is one in which researchers and participants are the predominant actors. Owing to a perceived absence of benefit to participants, courts have created a highly exacting duty to inform. Individual autonomy, as its name portends, is focused solely on the individual: in the case of biobanks and other research projects, that individual is the participant. When applying individual autonomy, the primary concern is with the actions researchers are required to take relative to participants, minimizing, at the same time, both the existence and interests of other actors.

The analysis in this chapter demonstrates that this understanding of the research relationship is untenable in the context of population biobanks. I focus on the concept of benefit by briefly exploring how the interests of the public have become central in benefit considerations. For population biobanks specifically, I rely on consent documents collected in Chapter 2 to demonstrate how the issue of participant benefit is portrayed during the consent process. I will then conduct a review of international, regional, and national documents retrieved from the PopGen database as a way of determining how researchers in population biobanks realize such benefit. This final exercise will anchor the important role of another actor, one that has thus far largely evaded consideration: the research community.

4.2.1 **The evolution of the concept of benefit**

According to the Oxford English Dictionary, "to benefit" is to "bring advantage to," or to "help or be useful to" [7]. The concept of benefit in medical research has received a wide array of interpretations in past decades. While certain authors link the concept to financial benefit [8], others associate benefit with therapeutic intent [9]. The concept of benefit in medical research ethics can be traced to the *Belmont Report,* a 1979 document adopted by the United States National Commission for the Protection of Human Subjects of Biomedical and Behavioral Research in the wake of the infamous Tuskegee syphilis experiments [10]. In the *Report,* "benefit" is defined as "something of positive value related to health or welfare" [10] that can "affect the individual subjects, the families of individual subjects, and society at large (or special groups of subjects in society)" [10]. The extent to which individual subjects may receive or extend benefit remains an important consideration in most kinds of research. In pediatric studies or research on incapable adults, for example, article 3.9 of the TCPS 2 states that a research ethics board should require the researcher to ascertain that:

> *the research is being carried out for the participant's direct benefit, or for the benefit of other persons in the same category. If the research does not have the potential for direct benefit to the participant but only for the benefit of the other persons in the same category, the researcher shall demonstrate that the research*

will expose the participant to only a minimal risk and minimal burden, and demonstrate how the participant's welfare will be protected throughout the participation in research [11].

Similarly, Article 21 of the *Civil Code of Quebec* [12] emphasizes the importance of benefitting minors and adults incapable of giving consent. An individual who falls under one of these categories

may participate in such research only if, where he is the only subject of the research, it has the potential to produce benefit to his health or only if, in the case of research on a group, it has the potential to produce results capable of conferring benefit to other persons in the same age category or having the same disease or handicap [12].

As can be seen in these examples, direct benefit that emanates from medical research may be associated with participants and other individuals within a particular age category or with those suffering from a specific disease or condition. This view entails that no one other than the participant (or someone in the same age category or having the same disease or handicap) will benefit from participation in the study. Of course, the abovementioned articles from the Civil Code of Quebec or the TCPS 2 cannot be broadly applied, for they only concern minors and incapable adults. With that said, they do demonstrate how considerations of benefit, at least as far as these vulnerable populations are concerned, remain largely focused on or modeled around the participants in question.

Canadian court decisions on the duty to inform in research diverge slightly from the examples mentioned above. In these decisions, consideration of benefit is thought to focus solely on the participant in question [13]. In other words, when benefit is an issue, the sole relevant determinant is whether the participant will benefit or not. There is no consideration for other individuals within the same age category or individuals suffering from the same disease, and even less still other unaffected individuals. Following from this individualistic view, it is "the absence of any therapeutic benefit to the patient which provides the policy justification for having different requirements for consent to research than for consent to treatment" [14]. The requirements associated with the duty of a researcher to inform became, as a result, more exacting. Nowhere in these decisions or in their subsequent interpretation by scholars was there a sense that the concept of benefit as understood by the court extended to stakeholders other than research participants.

Furthermore, reliance on considerations of individual benefit to delineate consent standards for invasive clinical trials that require the constant physical presence of participants [15] is unlikely to be useful in other kinds of research. There is no strong reason to think that a one-size approach is appropriate across research methods. Indeed, courts have yet to consider other, more observational and less individually centered research. For the time being, the duty to inform (and its correlative individualistic conception of autonomy) is framed in terms of a notion of benefit that conceives of the individual as the predominant actor in research. Given that research is generally understood to be principally focused on the production of generalizable knowledge, this is surprising on its face [16]. In fact, with the emergence of genomic research and biobanking, "generalizable" research has increasingly been associated with "populations" rather than "individuals":

> [i]nvestigators [...] are not expected to act primarily for the benefit of individual research participants, and indeed, should not if doing so might interfere with their ability to create generalizable knowledge [...] [16].

Population biobanks exemplify research projects in which direct benefit to individuals is not typically expected [17]. As a result, they offer an important example of research practices that are not reflected in case law. This reveals a clear conceptual shortcoming of the individualistic conception of autonomy. Before coming to understand the concept of benefit in population biobanks, however, we must first address debates that have developed in the human genetics context regarding benefit, given that this type of research has long been associated with biobanking [18].

In considering the concept of benefit, it is worth noting that reflections in the human genetics research context have long centered on the notion of benefit *sharing* [17,19−21]. First presented in 1992 by the *Rio Convention on Biodiversity* [22], "benefit sharing" referred to the just and equitable sharing of benefits derived from the use of genetic resources [17,23]. Although this convention focused on animals and plants, it inspired important discussions about the place of benefit in genetics research and on efforts to counterbalance the effects of commercialization for financially induced research participants [20]. In its 1996 *Statement on the Principled Conduct of Genetic Research* [24], the Human Genome Organization (HUGO) examined this problem and recommended

> that undue inducement through compensation for individual participants, families, and populations should be prohibited. This prohibition, however, does not include agreements with individuals, families, groups, communities, or populations that foresee technology transfer, local training, joint ventures, provision of health care or of information infrastructures, and reimbursement of costs, of the possible use of a percentage of any royalties for humanitarian purposes [24].

Of the seven benefits mentioned in the *Recommendation*, only one applies to individuals: the provision of health care. All other benefits bypass individuals and apply to other stakeholders. The view that benefit sharing, when achieved, should transcend the individual has been a central consideration in subsequent ethics norms. The UNESCO *Declaration on Human Genome and Human Rights*, for example, states that benefits generated by advances in research on the human genome should be made available to everyone [25]. Similarly, the HGO's Ethics Committee adopted a *Statement on Benefit-Sharing* in 2000 [26]. The Statement recommends that all of humanity should share in and have access to the benefits of genetics research [27,28] and that benefits should "not be limited to those individuals who participated in such research" [29]. In the same vein, but with more precision, UNESCO's 2003 *International Declaration on Human Genetic Data* [30] included a number of considerations for population-based studies and recommended that "benefits resulting from the use of human genetic data, human proteomic data, or biological samples collected for medical and scientific research should be shared with the society as a whole and the international community" [31]. The *Declaration* adds the following:

> *In giving effect to this principle, benefits may take any of the following forms:*
>
> a. *special assistance to the persons and groups that have taken part in the research;*
> b. *access to medical care;*
> c. *provision of new diagnostics, facilities for new treatments, or drugs stemming from the research;*
> d. *support for health services;*
> e. *capacity-building facilities for research purposes;*
> f. *development and strengthening of the capacity of developing countries to collect and process human genetic data, taking into consideration their specific problems;*
> g. *any other form consistent with the principles set out in this Declaration [31].*

In 2011, the *Nagoya Protocol on Access to Genetic Resources and the Fair and Equitable Sharing of Benefits Arising from their Utilization* divided benefits that might arise from genetic resources into two categories: monetary and nonmonetary [32]. The monetary category includes, but is not limited to, "access fees" [33] and "payment of royalties" [33]. The nonmonetary category, in contrast, includes such benefits as "sharing of research and development results" [34], "contributions to the local economy" [34], and "food and livelihood security benefits" [34]. It should be noted that, like most of what is listed in the Annex of the *Protocol*, these examples are not thought to be benefits targeted at specific participants.

Authors have increasingly been interpreting benefit sharing in terms of mechanisms that are put into place "to ensure that the benefits stemming from genomic research profit whole population groups [...]" [35] and away from individualistic calculations. Some authors have attributed this shift to increased attention to the concept of justice, writing that benefit sharing is "the action of giving a portion of advantages/profits derived from the use of human genetic resources to the resource providers to achieve justice in exchange [19]. Other authors have been motivated by a principle of fairness [20]. As in the HUGO and UNESCO documents, a number of authors have described the shift by drawing on the concept of "reasonable availability" [21], asserting that reasonable availability "requires that research be tailored to the health needs of the host community and that research results thus be made available to the community at the end of the project" [21].

In the following section, I will explore the concept of "benefit" as it applies in large-scale population studies and explore whether a shift in focus away from the individual per se to society at large is realized in the information such projects share with their participants.

4.2.2 **How are participants informed about benefits during consent?**

In the section above, I described how genetics and genomics research have ushered in a novel interpretation of the concept of benefit. More specifically, large-scale genomics research, such as population biobanks, has advanced a conception of benefit based on entire groups and communities. In light of this, it is integral to consider whether Canadian population biobanks have taken a similar approach.

In my analysis of how population biobank researchers have presented the concept of benefit to their participants, I relied on the consent forms, information brochures, and frequently asked questions collected in Chapter 2. Each of these documents (n = 12) was reviewed for any passage describing benefit. Once again, these documents were chosen because they best reflect the extent of information provided to participants during the recruitment process.

In Table 4.1 below, I highlight passages drawn from these consent forms and information brochures. These passages detail the interpretations these projects have taken of the concept of benefit. This reality, notably, cannot be limited to population biobanks, but can likely be associated with other nontherapeutic research projects.

These consent clauses clearly demonstrate that Canadian population biobanks do not predict much in the way of direct benefit to their participants. Certain cohorts use categorical statements, such as "participation [...] will not bring any direct benefit" [39] or "you will not get any direct personal benefit" [41]. Others use less uncompromising language, for example: "will likely not provide you with any direct individual benefits" [37] or "is not expected to provide you with any direct individual benefits" [40]. More critically, the clauses appearing in Table 4.1 highlight that population biobanks tend to express that their work is primarily expected to benefit society and future generations. BC Generations, for example, informs participants that "health benefits from this research are likely to help future generations" [36]. Both the Tomorrow Project and Atlantic PATH use similar language. The Canadian Longitudinal Study on Aging and the Canadian Alliance specifically refer to "society" as the major predicted benefactor of results emanating from the project [41,43]. This is similar to the trend identified in a number of normative documents and in the literature. Clauses in consent forms and information brochures reflect that the place of society and future generations in benefit considerations has been cemented. This tendency thereby expands the research relationship, incorporating explicit considerations beyond the interests of individual participants.

4.2.3 **Realizing benefits: maximizing collaboration with the research community**

Sections 4.2.1 and 4.2.2 highlighted the increasingly central place of the public and society at large in the research ecosystem. In particular, they pointed to a shift in our interpretation of benefit, away from the individual, and toward society and future generations. Following this, I will consider how such benefit might be realized. Put as a question, how can researchers in population biobanks materialize the benefit

Table 4.1 Consent provisions addressing benefits from Canadian population biobanks.

Name of the cohort	Portion of the cohort documentation
BC Generations Project *(British Columbia)*	**Consent form (version 4.0—December 12, 2014)** "The BC Generations Project will contribute to a better understanding about the causes of cancer and other chronic diseases and the factors that influence health and illness among Canadians. Health benefits from this research are likely to help future generations […]" [36].
The Tomorrow Project *(Alberta)*	**Study booklet (version DS3010v2—May 2011)** "Participation in the *Tomorrow Project* will likely not provide you with any direct individual benefits. […] The results of the *Tomorrow Project* will mostly help future generations. This study will lead to a better understanding of the causes of cancer, and potentially some of the factors that influence health and illness in a large group of Canadians" [37].
Ontario Health Study *(Ontario)*	**OHS website FAQ** "Thousands of volunteers in other long-term studies have contributed to research results that have helped to develop strategies to prevent disease or to increase early detection and to make treatment more effective" [38].
CARTaGENE *(Quebec)*	**Information brochure with consent form (April 7, 2014)** "Participation in CARTaGENE will not bring any direct benefit to the participant. However, studies conducted using CARTaGENE data and samples may lead to better medical knowledge and in turn improved health care" [39].
Atlantic PATH *(Atlantic Provinces)*	**Consent and brochure (version 9.2—March 6, 2013)** "Participation in this study is not expected to provide you with any direct individual benefits. […] The most important health benefits from the PATH study will be realized many years from now, and will largely help future generations. It will contribute to a better understanding of the causes of disease, and the factors that influence health and illness among a large group of Canadians" [40].
Canadian Longitudinal Study on Aging *(Canada)*	**Study information package—home interview and data collection site visit** "You will not get any direct personal benefit from taking part in the CLSA. It is possible that, someday, data and samples collected by the CLSA will lead to new tests that could help society, for example, a diagnostic test. Should this be the case, you will receive no financial gain" [41]. **Consent form—home interview and data collection site visit** "I understand that my information and samples will be used for research purposes only and this research may also have commercial uses that benefit society" [42]
Canadian Alliance for Healthy Hearts and Minds—Thunder Bay Site	**CAHHM—participant information and consent sheet (Thunder Bay Site)** "The Alliance project could provide society with a better understanding of the causes of chronic diseases and their risk factors. You are not expected to receive any direct medical benefit from your taking part in this study" [43].

to society that they have promised participants? The document analysis presented in the following section will outline how policymakers portray the realization of benefit in biobanking and the mechanisms required for its materialization.

4.2.3.1 Methodology

In reviewing the ways in which international, regional, and Canadian documents portray the realization of benefit in population biobanks, I relied on a document analysis using the PopGen Module [44] of the HumGen International Database [45], a database of guidelines and policies specific to human genetics research. Statements, Recommendations, and other similar documents were selected as sources in this work. Documents of a more binding nature, such as legislation, regulation, and enforced guidelines, were included to the extent they were available. As mentioned in my description of the methodology used in Chapters 2 and 3, the PopGen module categorizes documents according to levels of jurisdiction: international, regional, and national. Given the large number of results at the national level (more than 200 documents), my review of documents in the latter category focused on Canadian documents, simply because they are most pertinent to the focus of the book.

A total of 24 normative documents, consisting of mostly Guidelines, Statements, and Recommendations, were returned by the PopGen Module of the HumGen International database. Two legislative documents were also found. These documents were retrieved using *FULL TEXT* keywords such as "access AND sharing," in combination with the fixed *KEYWORD* "biobank." I chose not to add the term "benefit" to ensure that I have the chance to interpret documents holistically and not be limited to those that simply invoke the term "benefit" literally. I intended to leave options open in the case of documents that refer to benefit using an alternative designation. The search date range was established from 1990 to 2017, 1990 being the default set range of the PopGen search engine. As mentioned above, the PopGen search engine was discontinued following my initial search in 2017. To account for significant developments in the field, I conducted an updated search using publicly available online resources in 2021, using the search terms outlined above. I further cross-referenced included documents against the United States Department of Health and Human Services (HHS) *International Compilation of Human Research Standards*, a list of over 1000 laws, regulations, and guidelines enacted internationally and in countries around the world [46]. The HHS compilation includes general research standards, genetics research standards, and governance documents for human biological materials. The 24 documents initially returned were screened for pertinence in the biobanking field. To allow for as wide a perspective as possible, I retained results not necessarily specific to population biobanks, but relevant to biobanking generally. As a result of this screening, six documents were excluded. The remaining documents were reviewed in order to ensure that each addressed issues of benefit. Of the remaining 18 documents, 16 were selected for further appraisal.

One of the analyzed documents was legislative. Most were ethics norms emanating from international, regional, or Canadian organizations (n = 15).

Following a comprehensive assessment of retained documents, all were found to be complete—that is, they in fact considered the topic of benefit.

My analysis of these documents is relevant because consent forms, while important for understanding the dynamics of research consent and practice, often only allude to the question of benefit broadly. Moreover, they only represent the position of the population biobank itself. The comparative review performed in this section, however, incorporates a wider perspective, one that will help us understand how policymakers and international, regional, and Canadian expert organizations portray the realization of the concept of "benefit" in biobanking. The results of this analysis are described in the following three sections: (1) International Documents, (2) Regional Documents, and (3) Canadian Documents.

4.2.3.1.1 International documents[1]

As early as 1996, the *Bermuda Principles* [47] recommended that all human genomic sequencing information be made freely available in the public domain "in order to encourage research and development and to maximize its benefit to society" [47]. In 1998, the "HUGO's" Ethics Committee's *Statement on DNA Sampling: Control and Access* [26] stated that research samples obtained with consent may be used for other research if "there is general notification of such a policy, the participant has not objected, and the sample to be used by the researcher has been coded or anonymized" [26]. The statement also highlights that advances stemming from "other research" should benefit the general population for disease prevention and treatment [26]. In 2002, the HUGO Ethics Committee's *Statement on Human Genomic Databases* [28] supported this view by stating that "[i]nsofar as it benefits humanity, the free flow, access, and exchange of data are essential" [48]. It is worth noting that the exchange of data here refers to the access of data and samples by the research community.

The *Bermuda Principles* were revisited in the 2003 *Fort Lauderdale Rules* [49], which recognized that "the scientific community will best be served if the results of community resource projects are made immediately available to free and unrestricted use by the scientific community to engage in the full range of opportunities for creative science" [49]. Community projects, such as population biobanks, were defined as research projects "specifically devised and implemented to create a set of data, reagents, or other material whose primary utility will be as a resource for the broad scientific community" [49]. These principles have been reaffirmed in other normative statements, most notably in the 2008 *Amsterdam Principles* [50], which recommended expanding their application to other kinds of data, such as proteomic data (which refers to the set of proteins produced in an organism) [51]. The 2009

---

[1] Portions of this section have previously appeared in Zawati MH, There Will be Sharing: Population Biobanks, the Duty to Inform, and the Limitations of the Individualistic Conception of Autonomy. Health LJ. 2014; 21.

Toronto *Prepublication Data Sharing Statement* [52] similarly reiterated the value of sharing data for a wider group of stakeholders, including cohorts and tissue banks.

Certain guidelines have encouraged states to play a more proactive role, such as UNESCO's *International Declaration on Human Genetic Data* [31], which upheld the need to regulate, "in accordance with their domestic law and international agreements, the cross-border flow of human genetic data, human proteomic data, and biological samples so as to foster international medical and scientific cooperation and ensure fair access to these data" [53]. According to the *Declaration*, benefits resulting from the use of genetic and proteomic data should be shared with "the society as a whole and the international community" [31]. In October 2009, the OECD expressly addressed access issues in its *Guidelines on Human Biobanks and Genetic Research Databases* [54]. These *Guidelines* proposed that biobankers should, in order to advance knowledge and understanding, strive to make data and samples widely available to the research community [55]. The international Global Alliance for Genomics and Health's 2014 *Framework for Responsible Sharing of Genomic and Health-Related Data*, moreover, listed the "development of new scientific knowledge and applications, enhanced efficiency, reproducibility and safety of research projects or processes, and more informed decisions about health care" [56] as potential benefits of data sharing.

4.2.3.1.2 Regional documents[2]

At the regional level, access to data and samples has been addressed by the European Society of Human Genetics. Its recommendations on *Data Storage and DNA Banking for Biomedical Research: Technical, Social, and Ethical Issues* [57] claim there is an ethical imperative to promote access and the exchange of information, so long as confidentiality is protected. Indeed, Recommendation 17 states that "the value of a collection is proportional to the amount and quality of the information attached to it. The full benefits for which the subjects gave their samples will be realized through maximizing collaborative high-quality research."

Taking a similar position, the Council of Europe's Recommendation Rec (2006) 4 *of the Committee of Ministers to member states on research on biological materials of human origin* encourages the transborder flow of biological material and associated data where recipient states can ensure adequate levels of confidentiality protection. The Recommendation further affirms that member states should take steps to facilitate researcher access to data and samples stored in population biobanks [58]. Broadly, the Recommendation takes the view that such use and collaboration will contribute to improving the quality of life [59]. As a consequence, the European Commission recommended in its 2012 report *Biobanks for Europe—A Challenge for Governance* [60] that "greater investment should be made in the

---

[2] Portions of this section have previously appeared in Zawati MH, There Will be Sharing: Population Biobanks, the Duty to Inform, and the Limitations of the Individualistic Conception of Autonomy. Health LJ. 2014; 21.

development of e-governance tools to embed 'ELSI [ethical, legal, and social issues] by design' solutions, which can be used to augment existing governance structures and facilitate the sharing of samples and information between biobanks and researchers at a metalevel" [61].

4.2.3.1.3 Canadian documents[3]

Guidelines applicable in Canada tend to only partially address access and international research collaboration. Health Canada's 2011 guidance on *Biobanking of Human Biological Materials* [62], for example, stresses the importance of handling access requests in a timely manner in order to facilitate research activity [62].

Likewise, the TCPS 2 addresses genetic research on communities and includes a chapter on *Human Biological Materials Including Materials Related to Human Reproduction*, which underscores the importance of access and collaboration between researchers. It highlights that:

> Access to stored human biological materials—and associated information about individuals whose materials are banked—can be particularly useful in helping researchers understand diseases that result from complex interactions between our genetic makeup, environmental exposure, and lifestyles [63].

Along similar lines, Quebec's Network of Applied Genetic Medicine issued a *Statement of Principles on the Ethical Conduct of Human Genetic Research Involving Populations* in 2000. This document, subject to a number of conditions, promotes open access to biobanks under the principle of freedom of research [64]. Moreover, the Statement promotes collaboration between foreign researchers and the dissemination of research results in the explicit aim of contributing to the welfare of humanity [64].

4.2.3.1.4 Conclusion

These international, regional, and Canadian instruments appear to reflect one of the key characteristics of population biobanks: that they make data and samples available for future research. More importantly, they clearly point to the research community as an actor of fundamental importance, one that has not, thus far, been considered by the individualistic conception of autonomy.

In fact, the documents examined above each imply that data and samples should be shared with the broader research community in order to facilitate scientific advancement and to maximize benefits derived from the participation of individuals. These benefits, of course, extend to the population at large. This position corresponds to the approach taken by research funders when considering collaboration. Indeed, in 2011, a joint statement on *Sharing Research Data to Improve Public*

---

[3] Portions of this section have previously appeared in Zawati MH, There Will be Sharing: Population Biobanks, the Duty to Inform, and the Limitations of the Individualistic Conception of Autonomy. Health LJ. 2014; 21.

Health [65] led by the UK Wellcome Trust called for the equitable, ethical, and efficient sharing of data as a way of accelerating improvements in public health. The joint statement has since been signed by 19 funders, including the Canadian Institutes for Health Research [66]. It contains principles in concurrence with those highlighted in the international, regional, and Canadian norms reviewed earlier. For example, the joint statement calls for efficiency in a way that echoes the position of the Global Alliance for Genomics and Health, which presented enhanced efficiency, reproducibility, and safety as some of the potential benefits of data sharing [66]. Further, the joint statement defines ethical access as that which "should protect the privacy of individuals and the dignity of communities, while simultaneously respecting the imperative to improve public health through the most productive use of data" [66]. This is quite similar to the principle outlined by the European Society for Human Genetics, which emphasized an ethical imperative to promote access and exchange of information, as long as confidentiality of participants is protected [57]. The UK Wellcome Trust—led statement goes on to say that there is a need to "ensure that research outputs are used to maximize knowledge and potential health benefits" [66] given that "the populations who participate in the research [...] have the right to expect that every last ounce of knowledge will be wrung from the research" [66].

In brief, this section has sought to show how policymakers conceive of the realization of benefit in biobanks. The answer, it appears, is that they understand benefit to society to be realized by maximizing collaboration with the research community. Indeed, in order to achieve the statistical significance necessary for investigations of gene—gene, gene—disease, and gene—environment interactions over time, large numbers of samples and data are required [67]. Only the supply of data and samples through collaboration between biobanks and researchers can achieve this requisite breadth.

In this section, I aimed both to highlight the consistent presence of society in benefit considerations and to understand the importance of collaboration between researchers and biobanks as a means of promoting benefit to society. This analysis has stressed the limitations of the individualistic conception of autonomy in understanding the researcher—participant relationship in a way that is restricted to only these two stakeholders. It has shown how society and the research community play a similarly important role in benefit considerations. By extension, both society and the research community ought to become more important considerations when disclosing information to participants during the consent process. In fact, they, along with the population biobank and the participant, function within an interconnected web of relations. This is the issue for discussion in the following section.

4.2.4 Maintaining the Dynamic

I argue that the individualistic conception of autonomy does not place sufficient emphasis on the role of either society or the research community when disclosing information during the consent process. That being said, we might question why it is important to incorporate society and the research community. The answer is that all four stakeholders—participant, population biobank (researcher), society, and the research community—are part of a relational dynamic that must be maintained if population biobanking is to succeed [68]. While the goal of clinical care should be to provide a direct benefit to patients, population biobanks—as seen in Section 4.2.2— primarily aim to benefit society or a particular subpopulation (in rare disease research, for example). To realize these goals, however, mechanisms facilitating collaboration with the research community need to be in place. This was made evident in Section 4.2.3. By participating in a population biobank study, research participants are contributing data and samples for future, unspecified research. Once these data and samples are stored, biobanks often have an obligation to make them available to the research community. The goal is to increase the statistical power needed to generate useful results, which, in turn, will translate into a greater abundance of knowledge [54] for the benefit of society [47] and future generations. The ultimate goal is better population health and a corresponding increase in public trust when these better health outcomes materialize (see Fig. 4.1 below).

A narrow view of autonomy through liberal individualism devalues the potential influence of both society and the research community over the life of population studies. Individual autonomy does this by demanding the application of specific consent in the population biobank setting. As I described in Chapter 3, specific consent practices would require that participants explicitly reconsent to every access request submitted by a researcher. Were biobank researchers to follow such an approach, there is a realistic chance that the dynamic created between various stakeholders would be greatly destabilized.

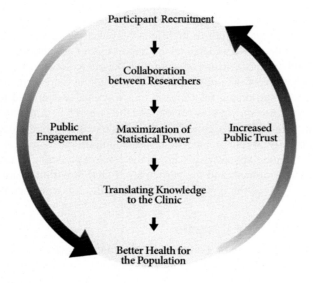

FIGURE 4.1

Maintaining the dynamic.

More specifically, consent requirements would create an overly complicated access system, one that would not respect what was promised to participants during the consent process: namely, that data and samples will be used for the benefit of future generations. In order to satisfy the requirement set out in Canadian case law that participants be informed of all facts and probabilities, biobank researchers will be obliged to reconsent participants every time a new research access request to their data and samples is made. Clauses presented in Table 2.2 of Chapter 2 show how population biobanks are clearly limited in what they are capable of divulging. Further, such reconsent would be far too cumbersome and costly for most projects, which usually involve more than 10,000 individuals, to feasibly undertake [69]. More importantly, delays would be incurred by members of the research community, who might choose not to make use of data and samples from a population biobank that has overly cumbersome procedures. Documents analyzed in Section 4.2.3 of this chapter have called for researchers to "maximize collaborative high-quality research" [57], recommending that they also "facilitate the sharing of samples and information" [61] by making them "immediately available" [70]. Requiring that researchers wait until the biobank is able to reach a participant and ask whether they wish that their data and samples be used by the applicant is not productive. Researchers would incur delays contacting participants (if it is even still possible to do so) and confirming their preferences. This process, based on individualistic concerns, would also undermine the very nature of population biobanks, which includes the creation of a governance system for organizing access to data and samples in a way that protects the interest of participants and sustains public trust. Indeed, as we saw in Chapter 2, Canadian population biobanks have put in place a governance system to ensure that access is carried out in a way that respects the wishes of participants (as expressed in their consent forms) and protects their privacy and confidentiality [71]. Reconsenting participants would ignore such governance and render it largely irrelevant. More importantly, by impeding the sharing of data and samples through reconsent, an individualistic conception of autonomy also risks hampering the return of enriched data emanating from the use of the data and samples by researchers [72]. This would ultimately impede the orderly translation of knowledge to the clinic [1] and by extension, to society as a whole. While participants do not expect any direct benefit from their participation, they expect that their data and samples will be used in an orderly fashion to advance science and produce generalizable benefits for future generations. Impeding this translational mechanism through unduly burdensome procedures based on individualistic concerns would risk sidelining the research community and fail to generate public benefit. Beyond that, a focus on individualistic concerns ultimately means that population biobanks will fail to respect what they promised to participants during the recruitment and consent phase. Ultimately, this reveals the relationship of interdependence that exists between research participants and other population biobank stakeholders. This dynamic is jeopardized when autonomy is seen as a form of independence precluding the research participant from considering the interests of other stakeholders in the multilateral relationships implicated in population biobanks.

The following section will examine several proposed solutions to the limitations created by individual autonomy. I will review them and examine whether they can be adaptable to population biobanks.

4.3 Proposed solutions and their limitations for population biobanks

A number of authors have offered models for mitigating the current status quo by palliating the shortcomings of "individual autonomy." As I mentioned in the introduction to this chapter, the individualistic conception of autonomy has received a fair deal of criticism. Certain authors have proposed new conceptions inspired by a range of theoretical currents. Most of the solutions, saved for the duty to participate in research, were proposed with the clinical setting in mind. None were offered as solutions to address the shortcomings of individual autonomy when disclosing information to participants during the consent process in population biobanks. I will briefly present the central tenets of these proposals and comment on whether they can conceivably be adapted to the population biobank setting. To do so, I will evaluate each model using the shortcomings identified for individual autonomy. Overall, four models will be examined: (1) deliberative model; (2) principled autonomy; (3) the duty to participate in research; and (4) relational autonomy. The deliberative model, principled autonomy, and relational autonomy all focus on the concept of autonomy, while the duty to participate in research proposes something substantially more expansive.

4.3.1 Deliberative model

In the early 1990s, Emanuel and Emanuel [73] described four models of the physician—patient relationship. The first was the paternalistic model, where a physician acts as the patient's guardian, prioritizing well-being over their free choice [73]. The second is the informative model, which, on Emanuel and Emanuel's account, represents the current "individualistic" model used in contemporary bioethics and law. In this model, the objective of the physician—patient interaction is "for the physician to provide the patient with all relevant information, for the patient to select the medical intervention he or she wants, and for the physician to execute the selected interventions" [73]. Emanuel and Emanuel further suggest that this conception embodies "a defective conception of patient autonomy" [73], one that "reduces the physician's role to that of a technologist" [73]. It does so simply because it is a physician's duty to provide all information no matter the values espoused by the patient and without the "fabric of knowledge, understanding, teaching, and action" [73] embodying the essence of what it is to be a doctor. The authors also present the interpretive model, in which the physician—patient interaction aims at identifying the values of the patient and helping them to select the medical interventions that best reflect them [73]. Finally, they present the deliberative model, which, in contrast to the interpretive model, aims at helping the patient determine and choose "the best health-related values that can be realized in the clinical situation" [73]. The goal of the physician, then, would be to persuade the patient of the most estimable values worthy of being pursued [73]. The authors are quick to differentiate the deliberative model from paternalism, as the former applies persuasion, while the latter prefers imposition [73]. Although they mention that each of these models has a certain degree of merit and could be justifiable in some circumstances, Emanuel and Emanuel prefer the deliberative model.

If we were to adapt this conception of patient autonomy to participants in population biobanks, would it be capable of palliating the limitations identified for the individualistic conception of autonomy? The answer, I think, is no.

The deliberative model's conception of autonomy is one of moral self-development, in which "the patient is empowered not simply to follow unexamined preferences or examined values, but to consider, through dialogue, alternate health-related values, their worthiness, and their implications for treatment" [73]. This model incorporates two basic elements: (1) moral persuasion by the physician and (2) the implementation of the patient's selected intervention. Essentially, a physician will recommend a particular intervention and try to persuade the patient that it should be taken. Following this, it is left to the patient to make a final decision.

If we were to adapt the deliberative model in the participant—researcher relationship, it would be up to the researcher to persuade participants of the values that best encapsulate the nature of the research project, including suggesting why certain values are more worthy of being pursued than others. Can the researcher then suggest to the participant that the public and the research community are important stakeholders in need of special attention in decisions surrounding participation? The answer is no, unless the researcher clearly presents this as an issue affecting the participant directly. In fact, according to Emanuel and Emanuel's original description of the deliberative model in the clinical setting, the physician only discusses values that are *related to the patient's disease* and treatments within the scope of their relationship [73]. If we were to adapt this conception to the population biobank research setting, the scope of the relevant values would need to have a direct impact on the participant to the exclusion of others. An argument may be made that not taking the public or the researcher community's interests into consideration would affect the feasibility of what is promised by the population biobank to the research participant. That said, all of this depends on the researcher being aware of these facts and, beyond that, being interested enough to convey them to the participant. It also means that it is not a consistently inbuilt approach that could be followed irrespective of the researcher interacting with the participant. For these reasons, despite breaking the individualistic conception's unidirectional approach to disclosing information to participants by creating an iterative bilateral approach, it would be difficult to assert that the deliberative model comprehensively or consistently grounds considerations related to stakeholders outside the researcher—participant relationship, including those surrounding benefit to the general public.

4.3.2 **Principled autonomy**

Like Emanuel and Emanuel, Onora O'Neill, a British philosopher, criticizes the shortcomings of individual autonomy in bioethics. In her view, individual autonomy has become an inflated term for informed consent requirements and its purported priority over other principles in bioethics should be seen as illusory [74]. In her book *Autonomy and Trust in Bioethics*, she emphasizes the need to "identify more convincing patterns of ethical reasoning, and more convincing ways of choosing policies and action for medical practice and for dealing with advances in the life sciences and in biotechnology" [74].

Enter, "principled autonomy" [74]. O'Neill, coining this concept, writes that the goal of autonomy should mainly be to ensure that no one is coerced or deceived rather than to guarantee that autonomous choices are protected [74]. According to Kant's conception of autonomy, principled autonomy suggests that the wrongs informed consent aims to protect against again, such as coercion and deception, are wrongs in their own right, independent of a need to respect autonomous individual choices. For O'Neill, this is made clear in Kant's use of the language of "autonomy of reason," "autonomy of ethics," "autonomy of principles," and "autonomy of willing," rather than language that associates autonomy to individuals per se [74].

O'Neill writes that principled autonomy is "not relational, not graduated, not a form of self-expression; it is a matter of acting on certain sorts of principles, and specifically on principles of obligation" [74]. On her view, Kant's conception of autonomy is one of self-legislation, in which individuals are obliged to act according to ethical reasoning [74]. Ethical reasoning, in turn, is based on "the ideal of living by principles that at least could be principles or laws for all" [74]. In other words, the principled autonomy account is predominantly concerned with the universalizability of principles of conduct [3,75]. If we were to adopt this conception in the research setting, it would imply identifying shared moral principles, accepted by researchers and participants alike for which all stakeholders would generally trust the others to follow such principles. Consequently, according to principled autonomy, the information-giving process would aim at thwarting instances of exploitation, research misconduct, and coercion, for example, rather than on cursorily upholding personal autonomy [74]. The participant would trust that a researcher will abide by moral principles and not aim to harm or exploit them.

Would principled autonomy, then, represent an appropriate model for population biobank research? The answer, again, is no. Principled autonomy suggests a plausible way of breaking the unidirectional relationship between researcher and the participant when the provision of information during the consent stage is considered. Importantly, this is so only insofar as they both act according to established moral principles and that the researcher does not attempt to abuse or coerce the participant. However, given that principled autonomy is not relational in nature, everything depends on the shared universal moral principles followed by the researcher and the participant. While O'Neill highlights both the lack of coercion and deception as pillars of autonomy, other moral principles at the core of ethical reasoning could be included. However, like the deliberative model, the bonds that O'Neill is trying to solidify are fundamentally those between researchers and participants. Nowhere is it clear that the public, society, or the research community would be able to have a place in the equation (especially in benefit considerations), or that the provision of information by the researcher, or choices made by the participant, would be called to account for the interests of other actors. One could argue that the invocation to consider the interests of other stakeholders in the research relationship is a shared moral principle. However, much like my argument against the deliberative model, this will depend on the particularities of the researcher and participant in question. More precisely, it will depend on whether they abide by these principles as a way of expecting the other party in the relationship to follow them as well. In other words, principled autonomy does not provide a solid basis to comprehensively and consistently ground considerations related to stakeholders outside the researcher–participant relationship in the population biobanking setting. It cannot, therefore, be an appropriate solution to palliate the shortcomings of the individualistic conception of autonomy.

4.3.3 The "moral duty" to participate in research

In this section, I will briefly explore the concept of a "moral duty" to participate in research that has been proposed by several authors [76–80], but mainly John Harris and Sarah Chan. I then examine the adaptability of this concept to population biobank research.

The concept of moral duty to participate in scientific research is not based on autonomy. Despite this, its central tenet calls for individuals to contribute to social practices that benefit themselves individually, in their role as members of society [79]. According to its proponents, the moral duty to participate in research is justified by the following factors: (1) fairness and (2) the duty of beneficence. John Harris argues that the principle of fairness recognizes the importance of contributing to social practices that benefit individuals. Scientific research, on his view, produces benefits that individuals currently enjoy (such as advancement in vaccine development) and benefitting without giving back (being a "free rider") is unfair to the social institution [79]. In a later article, Chan and Harris explain this concept in greater depth:

> [if] you benefit from an institution or practice, such as the ongoing institution of scientific research, and accept the benefits that derive from that institution, then you have, in fairness, a reason to support the existence of that institution or participate in that practice [80].

According to Harris, the duty of beneficence [79] indicates that it is morally wrong to abstain from acting when we could otherwise prevent serious harm. In these cases, failing to act is morally equivalent to accepting responsibility for any harm that materializes [80]. According to Chan and Harris, "failing to prevent harm is as effective a way of ensuring that harm occurs, and hence as morally reprehensible, as doing harm directly" [80]. In this passage, the authors contend that participating in research is a way of preventing harm to others. Not to participate would, as a result, be morally equivalent to harming people directly. For Chan and Harris, this amounts to what they call a duty to rescue [80]. This duty is not limited to individuals in the future, but includes people in the present as well [80].

While a moral duty to participate in research is an interesting notion, one which recognizes the importance of stakeholders outside of the researcher—participant relationship (especially in benefit considerations), I believe, for several reasons, that it is not easily adaptable to population biobanks. First, it is hard to contend that a particular person's life or well-being now or in the future could certainly be *in peril* in the case that another does not participate in research. Being both observational and longitudinal in nature, participation in population biobanks does not provide any direct benefit to the participant, much less to anyone else in the short term. Over the long term, the relevant research is intended to produce generalizable knowledge in order to better understand disease etymology and ultimately enable better health outcomes. But we cannot always be sure that this goal will necessarily be realized in the sense that benefit to future members of society is not *assured*. As some of the documents reviewed in Section 4.2.2 of this chapter have shown, research in population biobanks "may lead to better medical knowledge and in turn improved health care" [39]. The key word in this context is "may," which indicates that such results are uncertain. The issue of certainty features greatly in Chan and Harris' argument and seems to be a pillar of their duty to rescue argument. Indeed, in their argument as to why the duty to rescue can apply to both future and existing individuals, Chan and Harris state that

> Intuitively, it seems correct that a duty to rescue X today is more pressing than one to rescue Y in a year's time. [...] If we could say with 100% certainty that without our intervention [i.e., our participation in research], X and Y would both suffer equal injury, but at different times, it is hard to see why our obligation to X is greater than that to Y [80].

The majority of population biobanks are longitudinal in nature and can sometimes span from 20 to 50 years in length [36,37,39,81]. A promise of "better medical knowledge"—as described to participants in consent forms—that could require decades to materialize, should not be conceived as an *assured* and *certain* way to prevent harm.

Second, what is perhaps more precarious in Chan and Harris' conception of beneficence is their suggestion that if some individual fails "to attempt a rescue that he could have effected; and in that case moral shame ought rightfully to attach to him in full measure, as it surely would to anyone who stands by in idleness when he could have saved a life" [80]. This line of argument is problematic. At the authors' own admission [79], research undertakings are still met with skepticism on the part of the general public. While Chan and Harris do not go so far as to advocate for compulsory participation, by using the language of "shame," they certainly advocate for a *responsibility* to participate that is incongruous with the fundamentally voluntary nature of research participation—a principle that traces back as far as the *Nuremberg Code* [82]. Even in the clinical setting, where interventions are made with therapeutic intent, and in which timeliness is often a critical variable, the concept of responsibility is generally frowned upon [83]. The absence of actionable evidence sometimes associated with certain domains of medicine is frequently described as a limitation of such an approach [84]. This lack of actionable evidence is increased in the population biobank context, where future benefits are hoped for, but will only materialize and be useful in practice if they have transitioned from research to the clinical setting [83]. For all of these reasons, the creation of a *responsibility* to participate in scientific research would be incongruous with the practice of population biobanking.

4.3.4 Relational autonomy

Relational autonomy, much like the deliberative model, principled autonomy, and the duty to participate in research, has been suggested as a potential response to the individualistic conception of autonomy [85]. In this section, I briefly outline relational autonomy and consider whether it may be adapted to the population biobanking context. More precisely, I will examine whether relational autonomy is capable of palliating the numerous shortcomings of individualistic autonomy identified in previous sections of this chapter.

Relational theorists argue that the traditional approach to autonomy is fundamentally anchored in liberal individualism [86]. Instead of shunning the resulting conception of autonomy, they aim to reconceptualize it in a manner that emphasizes social connectivity and interdependence [87,88]. This conceptualization was defended by Nedelsky in *Law's Relations*, in which she writes that "autonomy exists on a continuum. As we act (usually partially) autonomously, we are always in interaction with the relationships (intimate and social-structural) that enable our autonomy. Relations are then constitutive of autonomy rather than conditions for it" [88].

According to proponents of relational autonomy, individuals are socially embedded. It is impossible to conceive of them as fundamentally distinct from their connections to others [89]. In fact, identity is formed only within the context of social relationships that are, in turn, shaped by a complex array of intersecting social determinants [89,90]. The individualistic conception of autonomy, in contrast, ignores this proposed web of relations [90]. Relational autonomy operates on the assumption that decisions are not simply "ours" [90]. Those with whom we are in relation might play an important role in our decisions and will generally be affected by them [91].

Certain authors have distinguished causal and constitutive conceptions of relational autonomy [92]. According to the causal view, individuals "face external constraints [and in order] to exercise her autonomy, the individual must remain situated [in] relations; absent relations, she lacks autonomy" [92]. The constitutive view, in turn, suggests that individuals are directly constituted by their relations and their various concerns for others [92]. Both of these conceptions share a common characteristic: conceptions of autonomy must take into account external social conditions at some level [92]. In the population biobanking context, the causal view, in which individuals lose autonomy for failing to be situated within a web of relations, is less of a concern than issues raised by the constitutive view, which points to limitations in conceiving of a decision made by an individual—when they are completely separated from all connections—as a fully autonomous decision. For this reason, I focus the present examination of relational autonomy by drawing primarily on the latter proposal. For reasons of brevity, I will set aside the causal theory and simply apply the language of relational autonomy to stand in for the constitutive conception.

Unlike the deliberative model, principled autonomy, or the duty to participate in research, relational autonomy represents a potentially stable foundation on which to construct a conception of autonomy that is cognizant of the complex, ongoing, and multilateral relationships that shape population biobanking projects. In such projects, multilateral relationships are founded in interactions among a number of stakeholders, including the population biobank itself, research participants, the public at large, and the research community. Since relational autonomy proposes that others might play a central role in the decisions of research participants—and be affected by them in turn—relational autonomy provides a potentially plausible framework in which to comprehensively and consistently ground considerations related to stakeholders outside of researcher—participant relationships in the population biobank setting (including benefit considerations). As I indicated in earlier sections of this chapter, decisions made by research participants tangibly affect both the public and research community. Relational autonomy reframes discussions during the consent process—whether related to risks, benefits, or general purposes of the research project—allowing the interests of the public and research community to be considered while also encouraging the research participant to be more sensitive to these interactions.

That being said, two concerns might be foreseen. The first is a practical worry. While we can conceive of a place for relational autonomy in the population biobanking context in principle, how this conception will be translated into practice is an entirely different issue. As a recent paper puts it "whether this reconceptualization of autonomy [i.e., rational autonomy] is taken up in practice largely will depend on how we [...] conceive it and what we want it to do" [92]. In the population biobanking context, there is a general, pervasive absence of practical clarity. In order for such practical clarity to be realized, a discussion of the nature and characteristics of the relations that exist in population biobanking is required. This is so primarily because such characteristics are necessary for adapting relational autonomy in that particular context while taking into account all of the existing stakeholders. The second concern relates to the possible infringement of individual rights when employing a relational conception of autonomy that takes external players into account and considers external conditions. Some authors, for example, have expressed a worry that relational accounts may end up defeating autonomous choices [89,93]. For example, decisions by a pregnant woman may be disregarded in the interests of a future child [93].

While significant, these concerns may be palliated by situating relational autonomy in a conceptual framework that practically describes, acknowledges, and sustains the multilateral relationships implicated in population biobank research, without also compromising the rights of participants. With this in mind, I propose that the concept of reciprocity is capable of doing exactly this. Over the next two Chapters, I will outline the core tenets of reciprocity and examine how it may be practically applied in the population biobanking context.

4.4 Conclusion

Advances in medical research necessitate the creation of reference maps of whole and subpopulations. Such maps serve a crucial role "as controls for replication, comparison, and validation of personalized genomic discoveries and profiles" [94]. Population biobanks play precisely this role. They are at the center of these vital public health planning pursuits. The only way biobanks will be capable of achieving their objectives is through the collection, storage, and sharing of data and samples for future unspecified research.

However, in order to efficiently undertake these activities, population biobanks must ensure that local legal requirements are satisfied. Chapter 2 described the nature, role, and characteristics of population biobanks as essential resources for researchers. In Chapter 3, I stressed the limitations of jurisprudential requirements of disclosure, which affect both the content and manner in which information is provided to participants during consent. Chapter 4 critiqued the restrictive nature of the traditional conception of autonomy—which lies at the theoretical heart of the exacting legal duties Canadian law imposes on researchers. I then demonstrated how its origins are rooted in a unilateral conception of autonomy that does not cohere with considerations beyond the participant—researcher relationship. Because the realization of benefits to the public requires maximizing collaborations among researchers, this chapter has demonstrated how an individualistic conception of autonomy could impede that from happening.

This chapter is in no way a repudiation of the importance of autonomy. As one author puts it "[c]ritiques of autonomy should not be taken as suggestions to do away with it. Instead, we should seek principles to complement it, especially when autonomy falters or is inapplicable" [95]. In line with this approach, I mean only to suggest that the individualistic conception of autonomy faces important limitations in population biobanking. This pushes us to identify a conception, premised on multilateral trust and transparency, that acknowledges the critical roles played by the general public and research community. To that effect, four candidate alternatives were reviewed as possible alternatives that would address the shortcomings of individual autonomy. The deliberative model, principled autonomy, and the duty to participate in research were, in turn, shown to be similarly inimical to population biobanking. None of these theories fully recognizes the complexities of benefit considerations and the importance of consistently incorporating the interests of stakeholders outside the participant—researcher relationship. Relational autonomy, however, was identified as a potentially fertile grounding for a conception of autonomy I argue is more fitting in the case of the complex, ongoing, and multilateral relationships established by population biobanking projects. That said, I also argue that in order to reinforce its capacity to acknowledge and sustain multilateral relationships implicated in population biobank research, without infringing on the individual rights of research participants, relational autonomy must be complemented with the concept of reciprocity. In the next chapter, I introduce that very concept of reciprocity.

References

[1] Shabani M, Borry P. You want the right amount of oversight: interviews with data access committee members and experts on genomic data access. Genet Med 2016;18(9):892—7.

[2] Deschenes M, Sallée C. Accountability in population biobanking: comparative approaches. JL Med Ethics 2005;33(1):40—53.

[3] Laurie G. Genetic privacy: a challenge to medico-legal norms. Cambridge: Cambridge University Press; 2002.

[4] Stirrat GM, Gill R. Autonomy in medical ethics after O'Neill. J Med Ethics 2005;31(3):127—30.

[5] Chin JJ. Doctor-patient relationship: from medical paternalism to enhanced autonomy. Singap Med J 2002;43(3):152—5.

[6] Prainsack B. The "We" in the "Me": solidarity and health care in the era of personalized medicine. Sci Technol Hum Val 2018;43(1):21—44.

[7] Oxford English Dictionary. Online edition. https://en.oxforddictionaries.com/definition/benefit. [Accessed 21.03.15].

[8] Ganguli-Mitra A. Benefit-sharing and remuneration. In: Elger B, et al., editors. Ethical issues in governing biobanks: global perspectives. Aldershot: Ashgate; 2008. p. 217—29.

[9] Ross L. Phase I research and the meaning of direct benefit. J Pediatr 2006;149(1):S22.

[10] The National Commission for the Protection of Human Subjects of Biomedical and Behavioral Research. The Belmont report: ethical principles and guidelines for the protection of human subjects of research. Washington: US Government Printing Office; 1978. s. C, point 2.

[11] Canadian Institutes of Health Research, Natural Sciences and Engineering Research Council of Canada & Social Sciences and Humanities Research Council of Canada. How to address material incidental findings: guidance in applying (TCPS 2) (2018). Ottawa: Secretariat on Responsible Conduct of Research; 2019. Art 3.9.

[12] Civil Code of Québec. Article 21.

[13] Halushka v. University of Saskatchewan. 1965. 53 DLR (2d) 436 at 443—444, 52 WWR (ns) 608 (Sask CA).

[14] Robertson GB, Picard EI. Legal liability of doctors and hospitals in Canada. 5th ed. Toronto: Thomson Reuters Canada Limited; 2017.

[15] See e.g. US National Library of Medicine. Learn about clinical studies. 2017. https://www.clinicaltrials.gov/ct2/about-studies/learn#ClinicalTrials [Accessed 21.03.15].

[16] Clayton EW, McGuire AL. The legal risks of returning results of genomics research. Genet Med 2012;14(4):473–7.

[17] Knoppers BM. Population genetics and benefit sharing. Community Genet 2000;3(4): 212–4.

[18] Sheremeta L, Knoppers BM. Beyond the rhetoric: population genetics and benefit-sharing. In: Philips PWB, Onwuekwa CB, editors. Accessing and sharing the benefits of the genomics revolution. Dordrecht: Springer; 2007. p. 157–82.

[19] Schroeder D. Benefit sharing: it's time for a definition. J Med Ethics 2007;33(4):205–9.

[20] Simm K. Benefit-sharing: an inquiry regarding the meaning and limits of the concept in human genetic research. Genom Soc Pol 2005;1(2):29–34.

[21] Dauda B, Dierickx K. Benefit sharing: an exploration on the contextual discourse of a changing concept. BMC Med Ethics 2013;14(1):1–8. Avaiable from: http://www. biomedcentral.com/1472-6939/14/36.

[22] Convention on biological diversity. June 5 , 1992. 1760 UNTS 79 (entered into force 29 December 1993).

[23] Convention on biological diversity. June 5 , 1992. 1760 UNTS 79, Art. 15 (entered into force 29 December 1993).

[24] Human Genome Organization. Statement on the principled conduct of genetics research. 1996. http://www.eubios.info/HUGO.htm [Accessed 21.03.16].

[25] Universal Declaration on the Human Genome and Human Rights, UNESCOR, 29th sess, resolutions, item 16, 29 C/res. 31. 2005. art 5(c).

[26] Human Genome Organization (HUGO) Ethics Committee. Statement on benefit-sharing. 2000. https://www.who.int/genomics/elsi/regulatory_data/region/ international/043/en/ [Accessed 21.03.16].

[27] Human Genome Organization (HUGO) Ethics Committee. Statement on benefit-sharing. 2000. Recommendation 1, https://www.who.int/genomics/elsi/regulatory_ data/region/international/043/en/ [Accessed 21.03.16].

[28] Human Genome Organisation (HUGO) Ethics Committee. Statement on human genomic databases. 2003. http://www.eubios.info/HUGOHGD.htm [Accessed 21.03.24].

[29] Human Genome Organization (HUGO) Ethics Committee. Statement on benefit-sharing. 2000. Recommendation 2, https://www.who.int/genomics/elsi/regulatory_ data/region/international/043/en/ [Accessed 21.03.16].

[30] UNESCOR. International declaration on human genetic data, 32nd sess, resolutions, item 22, SHS/BIO/04/1 REV. 2003.

[31] UNESCOR. International declaration on human genetic data, 32nd sess, resolutions, art 19a SHS/BIO/04/1 REV. 2003.

[32] Secretariat of the convention on biological diversity, Nagoya Protocol on access to genetic resources and the fair and equitable sharing of benefits arising from their utilization (Montreal). 2011. https://www.cbd.int/abs/doc/protocol/nagoya-protocol-en.pdf [Accessed 21.03.16].

[33] Secretariat of the convention on biological diversity, Nagoya Protocol on access to genetic resources and the fair and equitable sharing of benefits arising from their utilization (Montreal), Annex 1. 2011. https://www.cbd.int/abs/doc/protocol/nagoya-protocol-en.pdf [Accessed 21.03.16].

[34] Secretariat of the convention on biological diversity, Nagoya Protocol on access to genetic resources and the fair and equitable sharing of benefits arising from their

utilization (Montreal), Annex 2. 2011. https://www.cbd.int/abs/doc/protocol/nagoya-protocol-en.pdf [Accessed 21.03.16].

[35] Joly Y, Allen C, Knoppers BM. Open access as benefit sharing? The example of publicly funded large-scale genomic databases. J Law Med Ethics 2012;40(1):143—7.

[36] BC Generations Project. Consent form. British Columbia; 2014. p. 3—5 (obtained through correspondence).

[37] The Tomorrow project. Study Booklet, Alberta; 2011. p. 4—6 (obtained through correspondence).

[38] Ontario health study. 2014. Website FAQ, https://www.ontariohealthstudy.ca/about-the-study/frequently-asked-questions/ [Accessed 21.03.11].

[39] CARTaGENE. Second wave information brochure for participants. 2014. https://cartagene.qc.ca/sites/default/files/documents/consent/cag_2e_vague_brochure_en_v3_7apr2014.pdf [Accessed 21.03.11].

[40] Atlantic PATH. Consent and brochure (obtained through correspondence). 2013. p. 2—4.

[41] Canadian Longitudinal Study on Aging. Study information package — home interview & data collection site visit. p. 2-9. https://www.clsa-elcv.ca/doc/414. [Accessed 21.03.17].

[42] Canadian Longitudinal Study on Aging. Consent form — home interview & data collection site visit. https://www.clsa-elcv.ca/doc/448. [Accessed 21.03.17].

[43] Canadian Alliance for Healthy Hearts and Minds. Participant information and consent sheet. [Thunder Bay Site] [obtained through correspondence].

[44] PopGen Module. International database on the legal and socio-ethical aspects of population genomics. 2017 [archived website], https://web.archive.org/web/20160315170944/www.popgen.info/home.

[45] See HumGen International. HumGen database: your resource in ethical, legal and social issues in human genetics. 2018. www.humgen.org [Accessed 21.03.16].

[46] Office for Human Research Protections. U.S. department of health and human services, international compilation of human research standards. Washington: Department of Health and Human Services; 2020.

[47] Human Genome Organisation (HUGO). Principles agreed at the first international strategy meeting on human genome sequencing. 1996. https://web.ornl.gov/sci/techresources/Human_Genome/research/bermuda.shtml#1 [Accessed 21.03.24].

[48] Human Genome Organisation (HUGO) Ethics Committee. Statement on human genomic database, principle 3.a. 2002. https://www.eubios.info/HUGOHGD.htm [Accessed 21.03.24].

[49] Human Genome Organisation (HUGO). Sharing data from large-scale biological research projects: a system of tripartite responsibility. Community Resource Project; 2003. www.genome.gov/Pages/Research/WellcomeReport0303.pdf [Accessed 21.03.16].

[50] Rodriguez H, Snyder M, Uhlén M, Andrews P, Beavis R, Borchers C, et al. Recommendations from the 2008 international summit on proteomics data release and sharing policy: the Amsterdam principles. J Proteome Res 2009;8(7):3689—92.

[51] EMBL-EBI. What is proteomics?. 2021. https://www.ebi.ac.uk/training/online/courses/proteomics-an-introduction/what-is-proteomics/ [Accesssed 21.03.24].

[52] Birney E, Hudson TJ, Green ED, Gunter C, Eddy S, Rogers J, et al. Prepublication data sharing. Nature 2009;461(7261):168—70.

[53] UNESCOR. International declaration on human genetic data, 32nd sess, resolutions, art 18 SHS/BIO/04/1 REV. 2003.

[54] OECD. Guidelines on human biobanks and genetic research databases, best practice 4.1. 2009. www.oecd.org/science/biotechnologypolicies/44054609.pdf [Accessed 21.03.11].

[55] OECD. Guidelines on human biobanks and genetic research databases, principle 1.C. 2009. www.oecd.org/science/biotechnologypolicies/44054609.pdf [Accessed 21.03.11].

[56] Global Alliance for Genomics and Health. Framework for the responsible sharing of genomic and health-related data. 2014. https://www.ga4gh.org/genomic-data-toolkit/regulatory-ethics-toolkit/framework-for-responsible-sharing-of-genomic-and-health-related-data/ [Accessed 21.03.24].

[57] European Society of Human Genetics. Data storage and DNA banking for biomedical research: technical, social and ethical issues. Recommendation 17 Eur J Hum Genet 2003;11(12).

[58] Council of Europe Committee of Ministers, Recommendation Rec. 4 of the Committee of Ministers to member states on research on biological materials of human origin, recommendation adopted 15 March 2006 (958th meeting of the Ministers' Deputies), art 20. 2006. https://wcd.coe.int/ViewDoc.jsp?id=977859 [Accessed 21.03.16].

[59] Council of Europe Committee of Ministers, Recommendation Rec. 4 of the Committee of Ministers to member states on research on biological materials of human origin, recommendation adopted 15 March 2006 (958th meeting of the Ministers' Deputies), Preamble. 2006. https://wcd.coe.int/ViewDoc.jsp?id=977859 [Accessed 21.03.16].

[60] EC. Biobanks for Europe: a challenge for governance. Report of the expert group on dealing with ethical and regulatory challenges of international biobank research Luxembourg. European Commission; 2012.

[61] European Commission Biobanks for Europe- A Challenge for Governance. Report of the expert group on dealing with ethical and regulatory challenges of international biobank research. Recommendation 7. Luxembourg: European Commission; 2012.

[62] Health Canada Panel on Research Ethics. Guidance for health Canada: biobanking of human biological material. 2011. s 2.8.2.4.

[63] Canadian Institutes of Health Research, Natural Sciences and Engineering Research Council of Canada & Social Sciences and Humanities Research Council of Canada. Tri-council policy statement: ethical conduct for research involving humans [TCPS 2]. Ottawa: Secretariat Responsible for the Conduct of Research; 2014. Chap. 12 D.

[64] Network of Applied Genetic Medicine (RMGA). Statement of principles: human genomic research. 2000. https://www.rmga.qc.ca/admin/cms/images/large/enoncedeprincipesrechercheengenomiquehumaine_en_000.pdf [Accessed 21.03.16].

[65] Walport M, Brest P. Sharing research data to improve public health. Lancet 2011; 377(9765):537−9.

[66] Wellcome Trust. Signatories to the joint statement. https://wellcome.org/what-we-do/our-work/sharing-research-data-improve-public-health-full-joint-statement-funders-health#the-joint-statement-of-purpose-5ea3. [Accessed 21.03.16].

[67] Burton PR, Hansell AL, Fortier I, Manolio TA, Khoury MJ, Little J, et al. Size matters: just how big is BIG? Quantifying realistic sample size requirements for human genome epidemiology. Int J Epidemiol 2009;38(1):263−73.

[68] McCullough LB, Cross AW. Respect for autonomy and medical paternalism reconsidered. Theor Med 1985;6(3):295−308.

[69] Tassé AM, Budin-Ljøsne I, Knoppers BM, Harris JR. Retrospective access to data: the ENGAGE consent experience. Eur J Hum Genet 2010;18(7):741—5.

[70] Human Genome Organisation (HUGO) Ethics Committee. Statement on human genomic database, community resource project. 2002. https://www.eubios.info/HUGOHGD.htm%3E [Accessed 21.03.24].

[71] Lemmens T, Austin LM. The end of individual control over health information: promoting fair information practices and the governance of biobank research. In: Kaye J, Stranger M, editors. Principles and practice in biobank governance. Farham: Ashgate; 2009. p. 243—51.

[72] CARTaGENE. Data and samples access policy, 8.2.1. 2018. https://https://cartagene.qc.ca/sites/default/files/documents/policies/Politique%20d%E2%80%99acc%C3%A8s%20CARTaGENE-CHUSJ_EN.pdf [Accessed 21.03.16].

[73] Emanuel EJ, Emanuel LL. Four models of the physician-patient relationship. J Am Med Assoc 1992;267(16):2221—6.

[74] O'Neill O. Autonomy and trust in bioethics. Cambridge: Cambridge University Press; 2002.

[75] Kant I. Groundwork for the metaphysics of morals [Zweig A, trans.]. In: Hill TE, editor. Oxford: Oxford University Press; 2009.

[76] Rhodes R. Rethinking research ethics. Am J Bioethics 2005;7:15.

[77] Forsberg JS, Hansson MG, Eriksson S. Why participating in (certain) scientific research is a moral duty. J Med Ethics 2014;40(5):325—8.

[78] Forsberg JS, Hansson MG, Eriksson S. Changing perspectives in biobank research: from individual rights to concerns about public health regarding the return of results. Eur J Hum Genet 2009;17(12):1544—9.

[79] Harris J. Scientific research is a moral duty. J Med Ethics 2005;31(4):242—8.

[80] Chan S, Harris J. Free riders and pious sons—Why science research remains obligatory. Bioethics 2009;23(3):161—71.

[81] Sak J, Pawlikowski J, Goniewicz M, Witt M. Population biobanking in selected European countries and proposed model for a Polish national DNA bank. J Appl Genet 2012; 53(2):159—65.

[82] Nuremberg Military Tribunals. Permissible medical experiments. In: Trials of war criminals before the Nuremberg military tribunals under control council law, vol 10:2. Washington, DC: US Government Printing Office; 1949.

[83] Caulfield T, Evans J, McGuire A, McCabe C, Bubela T, Cook-Deegan R, et al. Reflections on the cost of "low-cost" whole genome sequencing: framing the health policy debate. PLoS Biol 2013;11(11):3.

[84] Pray L. Personalized medicine: hope or hype. Nat Edu 2008;1(1):72.

[85] Ells C, Hunt MR, Chambers-Evans J. Relational autonomy as an essential component of patient-centered care. Int J Feminist Approach Bioethics 2011;4(2):79—101.

[86] Frank B. Réflexions éthiques sur la sauvegarde de l'autonomie. In: Barreau du Québec. Pouvoirs publics et protection. Cowansville: Yvon Blais; 2003. p. 183—99.

[87] Llewellyn J, Downie J. Relational theory and health law and policy. Spec Ed Health LJ; 2008. p. 193.

[88] Nedelsky J. Law's relations: a relational theory of self, autonomy, and law. Oxford: Oxford University Press; 2011.

[89] Christman J. Relational autonomy, liberal individualism, and the social constitution of selves. Philos Stud 2004;117(1/2):143—64.

[90] Herring J. Relational autonomy and family law. London: Springer; 2014.

[91] Walter JK, Ross LF. Relational autonomy: moving beyond the limits of isolated individualism. Pediatrics 2014;133:S16—23.

[92] Dove ES, Kelly SE, Lucivero F, Machirori M, Dheensa S, Prainsack B. Beyond individualism: is there a place for relational autonomy in clinical practice and research? Clin Ethics 2017;12(3):150—65.

[93] McLean SA. Autonomy, consent and the law. London: Routledge; 2009.

[94] Knoppers BM, Zawati MH, Kirby ES. Sampling populations of humans across the world: ELSI issues. Annu Rev Genomics Hum Genet 2012;13:395—413.

[95] Gessert CE. The problem with autonomy: an overemphasis on patient autonomy results in patients feeling abandoned and physicians feeling frustrated. Minn Med 2008;91(4): 40—2.

The concept of reciprocity: origins and key elements

5

5.1 Introduction

In the Introduction to this book, I described being interested in looking beyond the observable characteristics of the duty to inform of researchers in population bio-banking. To do so, I set out to examine the constitutive elements of the duty to inform, namely, the conception of autonomy that motivates how the duty to inform, and the relationships that follow from it, is interpreted by Canadian courts. In Chapter 1, I underlined the exacting nature of the duty to inform of researchers. I stressed that researchers are obliged to provide participants with all relevant facts, opinions, and probabilities related to a research project during the consent process. I demonstrated that the leading judicial interpretation of the duty to inform in Canada has individualistic autonomy, by way of liberal individualism, at its core. In Chapters 2–4, I outlined the multiple conceptual limitations that individualistic autonomy faces in the population biobanking context. It is so limited for two reasons. First, it fails to recognize the complexities of benefit considerations in the research setting. Second, given its unidirectional aims, individualistic autonomy does not acknowledge the multilateral relationships necessarily implicated in population biobanking research, including those that incorporate the broader research community and the general public. Following this, I demonstrated how various proposed solutions failed to resolve the shortcomings of individual autonomy in the context of population bio-banks. In doing so, I gave special attention to the alternative approaches of deliberative autonomy, principled autonomy, and the duty to participate in research. At the end of Chapter 4, I introduced the concept of relational autonomy and determined that, in principle, it could be adapted to the population biobank setting, especially when considering what to disclose to research participants during the consent process. To that end, I identified the work of a number of authors who have advocated for a conception of autonomy that would primarily turn on "relationships and social structures" [1]. With that said, I argued that relational autonomy may practically be adapted to the population biobank setting only to the extent that it is complemented by a framework that adequately reflects interactions between various stakeholders engaged in population biobank research, notably participants, the population biobank itself, the general public, and the research community.

Enter: reciprocity. Presented as an emerging concept in bioethics [2,3], reciprocity is based on the premise that individuals will "help or benefit others at least in part because [they] have received, will receive, or stand to receive beneficial assistance from them" [4]. The concept of reciprocity can be traced at least as far back as Cicero, who noted that "there is no duty more indispensable than that of returning a kindness" [5]. He adds that "all men distrust one forgetful of a benefit" [5]. More recently, reciprocity has been described as a form of mutuality [6], as a relationship that recognizes the essence of humanity [7] or the relational [8] alliance [9] between two persons. It has been similarly characterized as a principle vital to securing a society's success, "a key intervening variable through which shared social rules are enabled to yield social stability" [5]. The concept of reciprocity has been variously applied in such fields as social policy [10], economics [11], public health [12], and clinical health care [13]. It has, similarly, been the subject of much theoretical debate. A thorough understanding of the nature and constitutive elements of reciprocity, for example, appears in the work of American legal philosopher Lawrence C. Becker [14,15]. In his seminal book *Reciprocity*, Becker takes note of the wide range of materials written on the concept and laments that such diversity of view makes the development of a harmonized conception of reciprocity challenging [14].

Over the next two Chapters, I aim to demonstrate that the concept of reciprocity provides a plausible grounding for relational autonomy, a conception of autonomy that will need to be respected by researchers when disclosing information to participants during the consent process. More importantly, I argue that such reciprocity-based relational autonomy is the most fitting conception of autonomy in light of the many complex, ongoing, and multilateral relationships established by population biobank projects. In order to do so, I must first introduce the concept of reciprocity and its features. This will be the aim of this chapter. I begin by presenting key elements of the concept of reciprocity, namely the importance of having donors and recipients who undertake reciprocal exchanges with each other. I present these elements first because, regardless of the specific conception of reciprocity under examination, the presence of a donor and a recipient is a universally accepted condition. I will thereafter introduce two distinct conceptions of reciprocity: reciprocity for mutual benefit and reciprocity for mutual respect. For both of these conceptions, I will examine the criteria necessary for a reciprocal exchange to be categorized as one or the other. Finally, this chapter will assess the different characteristics of reciprocal exchange that are at the heart of the concept of reciprocity (see Table 5.1). I will do so by examining the *nature* of the reciprocal exchanges (whether they are individual or communal), the two major *scopes* of reciprocal exchanges (generalized and nonspecialized), the *flow* of the exchange (seriate or negotiated), and the *value* bestowed in the exchange (instrumental or symbolic). By understanding these different characteristics of the reciprocal exchange, I will be able to sketch out a theoretical basis on which the concept of reciprocity can be applied in the population biobank setting. This will be the work of Chapter 6.

Table 5.1 Reciprocity: concept, conceptions, and characteristics of the reciprocal exchanges.

Concept	Reciprocity
Conceptions	Reciprocity for mutual benefit *or* reciprocity for mutual respect
Nature of the reciprocal exchange	Individual *or* communal
Scope of reciprocal exchange	Generalized *or* non-specialized
Flow of the reciprocal exchange	Seriate *or* negotiated
Value of the reciprocal exchange	Instrumental *or* symbolic

5.2 Key elements of the concept of reciprocity: donors and recipients

Reciprocity has historically been interpreted in a number of ways. In its simplest formulation, reciprocity can be understood to have both positive and negative expressions. A *positive* expression of reciprocity is the provision of a good in exchange for something received. *Negative* reciprocity, on the other hand, refers to the return of hostility for hostility incurred [14]. For the purposes of this book, I limit my examination of reciprocity to its positive expression which I will describe simply as reciprocity. I focus only on positive reciprocity because, in the subject matter at hand, hostility between the relevant parties is unlikely to arise.

The concept of reciprocity is hardly new. All of us, after all, have had kind gestures returned by our benefactors. Consider two office colleagues arriving to work in the morning. One opens the door for the other. Once inside, the second reciprocates and opens the door for the first. This is a classic example of reciprocity, exemplifying a relationship in which the second colleague is a *recipient* and the first is a *donor*. Also consider the relationship between two neighbors. The first struggles to carry a new set of furniture into her home. The second, seeing the difficulty encountered by the first, offers a helping hand. In recognition of his help, the neighbor offers coffee and some snacks. In this case, the first neighbor is a recipient, and the helping neighbor is a donor.

Neither recipient in these examples was under any sort of legal obligation to reciprocate. Why then might they have chosen to act in this way? In his book, Lawrence Becker undertakes his assessment of reciprocity with this same question: "what can there be, in the very act of giving a gift, that requires a commensurate return on the part of the recipient?" [14] According to his view, the act of reciprocating is a moral virtue [14]. The relevant virtue, critically, is the *recipient's*—not, as we might expect, the donor's. Using the examples above, reciprocity seeks to understand the act performed in return by the second office colleague and by the neighbor who struggled with her furniture. Becker clarifies this point by writing that

Reciprocity is a recipient's virtue. It is the way people ought to be disposed to respond to others. It says nothing about how people ought to behave, or feel, when they give a gift. Perhaps friends ought to give without thought of a return. But how should we receive gifts from our friends? Surely we should not respond to them with evil, or with indifference. And surely we should make our responses fitting and proportionate. That is reciprocity [14].

The nature of reciprocity on this view, that it must be fitting and proportionate, will be discussed further in sections to come. For now, it is important to note that reciprocal relationships must have both a donor and a recipient. More importantly still, the study of reciprocal relationships focuses on the recipient's actions rather than on the donor's. Because reciprocity definitionally involves some kind of return, my focus will be primarily on the contours and characteristics of the recipient's return rather than on the act of donation. With that in mind, I turn to the following questions. First, how many conceptions of reciprocity exist? Second, how might the exchange between a donor and a recipient, which is at the heart of reciprocity, be qualified? To begin answering these questions, I will first present two distinct conceptions of reciprocity: 1) reciprocity for mutual benefit and 2) reciprocity for mutual respect.

5.3 Two conceptions: reciprocity for mutual benefit and reciprocity for mutual respect

After reading some of the description provided in the previous section, one might be led to question how reciprocity is related to the Golden Rule found in Confucian, Talmudic, and New Testament writings [14]. While such association is understandable, it is imprecise. The Golden Rule is significantly broader in scope, for it concerns much more than simple exchanges between two persons, "it proposes a criterion for initiatives one might take: [d]o to others only what you would have them do to you" [14]. This meaningfully differs from reciprocity, which is implicated only after a recipient responds to the actions of a donor. As I mentioned, the concept of reciprocity mainly focuses on assessing an exchange between a recipient and donor, with primary emphasis on the recipient. The Golden Rule, on the other hand, is interested only in the donor and what they ought, preemptively, to do for others.

As I mentioned above, two different conceptions of reciprocity exist: reciprocity for mutual benefit and reciprocity for mutual respect. Reciprocity for mutual benefit—which is defended by both Becker and Alvin Gouldner, an American sociologist—conceives of reciprocity as an exchange aiming to mutually benefit both the donor and recipient. Reciprocity for mutual respect, proposed by Christie Hartley, a philosophy professor, claims that reciprocity primarily aims to achieve mutual respect between a donor and recipient. In this section, I will introduce these alternative purposes of reciprocal action in order to later assess how they each might palliate the shortcomings of individual autonomy when considering the disclosure of information to participants during the consent process.

5.3.1 Reciprocity for mutual benefit

To better understand the reciprocal relationship between a recipient and a donor, one must first understand the purpose of reciprocal exchange. For Becker, an exchange might be connected to prudence, self-interest, altruism, justice, or fairness, among other things [14]. Whatever the motivation, Becker's position is that an exchange between a donor and recipient within a reciprocal relationship ultimately aims at producing mutual benefit: "[w]e ought to be disposed to return good for the good we get from agents who are trying to produce benefits for us" [14]. For Becker, such mutual benefit is further directed at promoting social equilibrium [14], a concept explained at length in Gouldner's writing [5]. In order for a reciprocal exchange to realize the aim of mutual benefit, it must be sensitive to matters of fittingness and proportionality, two notions that I examine in the following subsections. Before turning to these criteria in detail, I should make a small clarification. In Becker's work, the terms "donor" and "recipient" can sometimes be used to describe the same person at different stages of the reciprocal relationship. This is logical. A donor undertakes an act of donation toward the recipient. The recipient then reciprocates by responding to the donor. For Becker, in this response, the recipient will then become the donor and the donor will become the recipient. In other words, the two individuals exchange these positions successively. That said, in order to eliminate any confusion, I will use the language of recipient and original donor separately, even if these roles may shift throughout the reciprocal exchange. In order to better understand the conception of reciprocity for mutual benefit, it is useful to examine the constitutive characteristics of any resulting reciprocal exchange associated with it: (1) fittingness and (2) proportionality.

5.3.1.1 Fittingness

For Becker, a reciprocal exchange is fitting, that is, fulfills the criteria of fittingness, if (1) what is returned by the recipient is an objective "good" for the donor and (2) is both perceived by the donor as such and understood to be in return for his act of donation [14].

The first requirement of fittingness is straightforward: the return should be considered a "good" for the donor. More precisely, one should not return "evil" for a received "good." If I (a donor) give up my seat in the bus to a man with a physical handicap (a recipient) and he responds by unnecessarily putting his bag on the only seat remaining on the bus, his return would likely not plausibly be considered a good.

This brings us to a second requirement that the return is *perceived* by the donor as both a good and a return, which underlines an element of subjectivity in the overall assessment. Determining whether this condition is met will require a case-by-case analysis, subject to the circumstances at hand [14]. This second requirement of fittingness is squarely in the eye of the beholder—in our case, in the eye of the donor (whose initial act of donation will be reciprocated) [14]. In his book, Becker uses the example of an anonymous blood donor to illustrate this point. Donated blood could save the life of a hospitalized little boy. For the sake of argument, suppose that the boy's parents succeed in tracking down the anonymous donor. Should they decide to reciprocate by thanking the donor or buying them a gift, one might objectively say that this return is a good. The identified donor, however, might not see it that way. In fact, they might see their reidentification as a breach of privacy, even if the intentions were noble. In his book, Becker uses this example to suggest that a more fitting return would have been for the recipient to donate to the blood bank in turn [14]. That said, how will we know that the donor will perceive this as a good and as a return if he or she is anonymous? According to Becker, in the absence of a clear way of assessing how the original donor perceives the return, we should presume that the act is a good from the donor's perspective given that they donated to the bank in the first place and wished to remain anonymous. While we are not interested by particulars of donation in assessments of reciprocal exchange, Becker presumes that a return that is identical to the initial donation qualifies as a good. The essential logic is that a donor who did not think donation is a good would not have donated in the first place. Furthermore, Becker sees the act of donating to the blood bank (and not, for example, to a charity) by the recipient as fitting because it is the most convenient return possible in the circumstances. For all of these reasons, and when compared to tracking down the donor, donating to the blood bank might more easily be seen as a fitting return. Interestingly, it seems that the personal knowledge of the donor that a return was made by the recipient is unnecessary in the circumstances, especially given the wishes of the donor to remain anonymous.

Fittingness is only one of two criteria that must be satisfied for an exchange to be considered reciprocal. The following section will describe the second criteria: proportionality.

5.3.1.2 Proportionality

Proportionality is the second criteria needed to determine whether an exchange fits within a framework of reciprocity for mutual benefit. If a return is fitting according to the description given above, one must then assess whether it is proportional. In doing so, we should assess whether the return was "equal to the good received" [14]. This requirement is justified because reciprocity for mutual benefit—as is implied by its name—ultimately aims at producing a *balanced* exchange of benefits [14]. For Becker, the best possible return is one of commensurate benefits with as little sacrifice as possible [14]. You open the door for me, I open the door for you: equal benefit with minimal sacrifice [14].

An obvious problem then lies in cases for which it is not possible to reciprocate with precisely equal benefit. Consider the example of a Good Samaritan who donates a large sum of money to a poor family. The poor family is clearly unable to return a commensurate benefit, for doing so means they would lose everything they have. As Gouldner suggests: "the demand for exact equality would place an impossible burden even on actors highly motivated to comply with the reciprocity norm [...]" [5]. In such cases, an equal *sacrifice*, for Becker, becomes the most satisfactory option. An equal sacrifice would "not compromise the ability of either party to make further exchanges" [14]. A proportional return would then be an equal sacrifice, proportional to the recipient's situation when compared to that of the donor. Gouldner adopts similar reasoning and notes that the "obligations imposed by the norm of reciprocity may vary with the status of the participants within a society" [5]. To be more precise, if the large sum of money represents 1% of the donor's savings, a return by the poor family that amounts to 1% of what they have could be seen as proportional as it represents a roughly equal sacrifice.

In summary, a reciprocal act necessarily includes both a donor and a recipient. Reciprocity is not concerned with how donors should or should not act. It is concerned only with the actions of recipients. In order to be included in the conception of reciprocity for mutual benefit, an act must be both fitting and proportional. That said, reciprocity for mutual benefit is not the only proposed conception of reciprocity. The following section will explore reciprocity for mutual respect.

5.3.2 Reciprocity for mutual respect

Traditionally speaking, reciprocal relationships have been understood to aim at sustaining mutually advantageous relationships (reciprocity for mutual benefit). In recent years, however, that traditional conception has been increasingly challenged. Notably, criticism of reciprocity for mutual benefit features prominently in the work of Christie Hartley, a philosophy professor. For Hartley, reciprocity should not only be focused on mutuality of benefit but should also aim at showing "respect for someone who contributed to one's project" [16] and thereby be a form of recognition of a donor's contribution.

Hartley sets out her argument by invoking the example of a colleague who decides to stay late at the office to help a departing colleague pack up her things. In doing so, she misses a submission deadline for a conference paper [16]. The colleague who stays late to help (a donor) is doing so out of respect and kindness, but at a cost. The departing colleague (a recipient) shows her appreciation by gifting her a poster she had often found humorous. For Hartley, the colleague who sacrificed and stayed late will see the poster as a benefit, though receiving it could not compare to the way she benefited her departing colleague by staying late and missing a deadline. But the goal of this exchange, on Hartley's view, is not to sustain mutually beneficial relationships, but rather to thank the colleague, to show respect, and to recognize the contribution she made [16]. Hartley clarifies that although the exchange is asymmetrical, it is appropriate in the circumstances [16]. As a criterion that must be satisfied on this conception of reciprocity, Hartley proposes to retain fittingness. At the same time, she rejects proportionality and replaces it with a criterion of sufficiency, which aims to fit a new purpose: that of a reciprocal relationship based on mutual respect among equals [16].

In the following two subsections, I will present both Hartley's criterion of fittingness—which differs slightly from Becker's given the different purpose of the reciprocal relationship—and sufficiency.

5.3.2.1 Fittingness

For Hartley, fittingness, understood as a return that is objectively good for the original donor, remains an important component of reciprocity for mutual respect. Importantly, Hartley presents her conception of reciprocity within a framework that conceives of all individuals as free and equal cooperating members of society [16]. On her view, the aim of such cooperation is the creation and sustainment of "society based on relations of mutual respect among equals" [16]. Hartley defines equality, therefore, in terms of social relationships [16]. It is important to note that this conception of equality does not entail that people necessarily have equal responsibilities. Some of us could have greater responsibilities than others, while others, given their capacities, may have no responsibilities at all [16].

The requirement of fittingness, then, will be satisfied when a return contributes to members of society and does so in a manner understood to foster cooperation [16]. Expanding on the work of the political philosopher John Rawls [17], Hartley explains that social cooperation entails living among others on terms of mutual respect [16]. Thus, being in relationship with others in ways that contribute to a cooperative social structure and helping to produce goods needed by members of society may both be considered fitting contributions in the relevant sense [16]. The coworker who gave her colleague a gift in return for staying late did so as a way of acknowledging the contribution of the colleague and to show appreciation to another member of society through an act of mutual respect. The act, if fitting, satisfies the requirements of reciprocity for mutual respect so long as it also accords with the sufficiency requirement.

5.3.2.2 Sufficiency

Reciprocity for mutual benefit has the purpose of securing mutual advantage [14,16]. For this reason, the criterion of proportionality is critical in determining the appropriate balance of exchanges. In the case of reciprocity for mutual respect, however, Hartley proposes that proportionality is no longer necessary. It should, on her view, be replaced with a criterion of sufficiency. For Hartley, determining whether an act is sufficient is, in fact, closely related to the fittingness assessment. Indeed, if fittingness refers to a return that contributes to members of society in a manner understood to foster cooperation [16], sufficiency, in turn, requires that the return be fair in the sense that it is reasonably acceptable to rational individuals who aim to live and cooperate on grounds of reciprocity and mutual respect [16]. The criterion of proportionality found in reciprocity for mutual benefit requires that a return be quantified either as an equal benefit or an equal sacrifice. Contrastingly, sufficiency, according to Hartley, is not measurable because it is difficult to "quantify the value of relating to others in accordance with the substantive demands of a relationship based on mutual respect" [16].

In order to better understand sufficiency, Hartley provides the example of a donor who helps his neighbor (a recipient) to paint his house before it is sold. The neighbor wants to reciprocate. No matter what course the neighbor chooses, their act should not be seen as aiming to symmetrically balance the exchange, but should be sufficient to show respect as a cooperative contributor to the neighbor's project [16]. Obviously, the return should satisfy the fittingness requirement: a contribution understood to foster cooperation. Further, the return should recognize the contribution in a way that satisfies the criterion of sufficiency, namely, that it will be seen as fair by members of society. A good example of this would be for the neighbor to give the helper a souvenir, perhaps some trinket that they are very fond of. Returning the kind gesture with an item of that sort would satisfy the fittingness requirement. It is also sufficient because it contributes to another member of society in the sense that it exhibits mutual respect for others. It indicates recognition of a contribution made to the neighbor's project. It would be sensible to assume that rational individuals would see such return as one that would be reasonable to accept as fair.

One important thing can be drawn from this survey of the conception of reciprocity for mutual benefit and reciprocity for mutual respect is the central importance of the actual exchange. For this reason, it will be useful to understand the specific characteristics of reciprocal exchanges. If I am to apply both conceptions of reciprocity to population biobanking, examining the nature, scope, flow, and overall value of reciprocal exchanges is necessary.

Before delving deeper into these characteristics, it is important to note that the various conceptions of reciprocity I have outlined here (namely, reciprocity for mutual benefit or for mutual respect) may be realized in a range of reciprocal exchanges, sometimes combining vastly different characteristics. Put another way, neither conception has a predetermined set of characteristics (e.g., nature, scope, flow, and value) for reciprocal exchanges falling under its ambit. That said, some of these characteristics may be more easily associated with one conception than with the other. For example, a negotiated flow of reciprocal exchanges can be more easily associated with the conception of reciprocity as mutual benefit. However, the nature, scope, and value of that reciprocal exchange will not necessarily be the same for all cases of reciprocity for mutual benefit. I will refer to these situations in greater detail when examining the nature, scope, flow, and overall value of the reciprocal exchanges that can be applied to both reciprocity for mutual benefit and reciprocity for mutual respect.

5.4 Nature of the reciprocal exchanges

In previous sections, I established that a relationship would qualify as reciprocal only insofar as an exchange occurs. I now turn to examining the *nature* of reciprocal exchanges. In other words, who are such exchanges aimed at and how direct may they be?

To begin, two kinds of reciprocal exchange may occur: reciprocal exchanges that are individual in nature and those that are communal in nature [18]. Individual reciprocal exchanges concern one single individual. If my neighbor helps me to paint my house, returning his kind gesture by offering him a vase he so often praised is an individual reciprocal exchange in the sense that it concerns my neighbor and no one else. Communal reciprocal exchanges, in contrast, focus on a return to a potentially larger community of people. In the case of the blood bank example used in the earlier sections—in which the benefactor of an anonymous blood donation will return the kind gesture by supporting the blood bank—the return is communal in nature in the sense that it does not concern one specific individual, but potentially many of them.

Additionally, the first of these exchanges is direct. I have, in other words, *directly* returned something to my neighbor. According to Hobbs et al. individual reciprocal exchanges are *always* direct [18]. Likewise, communal exchanges are synonymous with indirect exchanges [18]. In fact, in the case of the blood bank illustration, when the original benefactor supports the blood bank, he or she completes an indirect reciprocal exchange given the return will not be directly provided to the original donor, but to future benefactors of the blood bank.

Beyond the direct or indirect nature of the exchange, there exist two major scopes of reciprocal exchanges: generalized reciprocal exchanges and nonspecialized reciprocal exchanges [19].

5.5 Scope of reciprocal exchanges

Generalized reciprocal exchanges focus less directly on monetary value. In exchanges of this kind, reciprocity occurs "primarily in terms of reputation, prestige, and power rather than in economic returns" [19]. In other words, the return in a generalized exchange likely will not be economically commensurate or even have monetary value at all. If a person donates a very large sum of money to a charity, the return will not be in the exact amount donated. The generalized reciprocal exchange will be completed through the reputation the donor will attain from giving such a large sum of money to a charitable cause. This is why, according to Macneil, generalized reciprocal exchanges are oriented toward maintaining social solidarity [19].

Macneil conceives of nonspecialized exchanges as perfectly balanced [19]. The default setting in cases of nonspecialized exchange is one of simultaneous exchange of identical goods. For example, I leave the door open for you when you come in the office, and you reciprocate by leaving the second door open for me. However, nonspecialized reciprocal exchanges may also include transactions that feature a commensurate return that is stipulated to obtain in some narrow window of time [19]. In other words, it is possible that the relevant exchanges will happen on a fixed timeline, rather than instantaneously, and will include commensurate rather than identical goods.

The examination of both generalized reciprocal exchanges and nonspecialized reciprocal exchanges opens the door to another discussion: one that is focused on the flow or reciprocal exchanges. More specifically, when will the reciprocal exchange be completed and can its completion be a condition of the original act of donation? The following section will discuss these issues in greater detail.

5.6 Flow of reciprocal exchanges

Reciprocal exchanges may flow either unilaterally or in a negotiated fashion. Unilateral exchange between individual actors refers to situations where each individual is free to initiate exchange with the other at any time. A unilateral flow entails that some initiations may be reciprocated immediately, while others will be reciprocated only later [20]. Put simply, when one initiates a reciprocal exchange, they should not expect to receive something in return immediately.

Despite the possibility of variation in the speed of return, any resulting exchange is nevertheless of a reciprocal nature, taking into account the circumstances at hand [20]. The key, then, is that an initial act of donation must not be *conditioned* on an immediate return. For this reason, I believe the term "seriate" is more appropriate when describing this type of flow. It includes all of the characteristics of a "unilateral" flow of reciprocal exchange but avoids being confused as being "unidirectional" or "one-sided." I use the word "seriate" in the sense that an initial act made by one person or entity toward another could occur at a later time and is not conditioned upon the latter reciprocating immediately to the former (see Fig. 5.1).

As for the negotiated flow of a reciprocal exchange, its name clearly explains its attributes. This is the opposite of a seriate flow in the sense that it refers to the completion of an agreement rather than an act that may or may not be reciprocated immediately. An agreement of this kind "creates a dyadic unit" [20] and specifies, as a transaction, what each party will receive from the other. This kind of exchange applies best to reciprocity for mutual benefit rather than reciprocity for mutual respect. Obviously, this type of exchange is conditional as "each actor's outcomes depend on the joint actions of self and other" [20] (see Fig. 5.1). The existence of an agreement

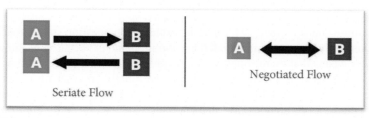

FIGURE 5.1

Flow of reciprocal exchanges.

does not mean that the return will be equal. All that is necessary in a negotiated flow is the agreement of the parties, whether or not the exchange is equal [20]. If I agree to pay for the installation of a stereo system in your car on the condition that you do the same when I buy mine, the flow of the exchange is negotiated. We might also agree that, in exchange of me paying for the installation, you will buy a stereo system for my house, which is likely to be more expensive. This is a negotiated flow all the same, even if the transaction is unequal.

Whether the flow of reciprocal exchange is seriate or negotiated does not necessarily affect their value. Put another way, there may not be a clear connection between the flow of exchange and the purpose of the exchange. In the next section, I will outline the possible value of reciprocal exchanges.

5.7 Value of reciprocal exchanges

Reciprocal exchanges may have instrumental or symbolic value [21]. The first of these is sometimes called utilitarian value, for it refers to acts of reciprocity that extend some form of utility to the recipient: "their value is instrumental in the sense that they help the recipient meet the need that was the original objective of the exchange" [21]. Here, as above, this characteristic applies best in cases of reciprocity for mutual benefit rather than in reciprocity for mutual respect. Indeed, in a reciprocal exchange that is instrumentally valuable, each party jointly receives negotiated returns that will help them realize their interests. Such reciprocal exchange may be linked to the negotiated flow discussed in the previous section. Given that the purpose of the exchange will have been previously identified and agreed upon, a negotiated flow will thereby likely enhance the instrumental value of the exchange.

Symbolically, valuable exchanges, in contrast, have been described as having value that is present in the reciprocal act itself and is neither instrumental nor derived [21]. Symbolic value has two constitutive elements: an "uncertainty reduction value" and an "expressive value" [21]. Uncertainty reduction refers to acts of reciprocity that "carry uncertainty reduction value to the extent that they reduce the risk and uncertainty inherent in exchange, by providing evidence of the partner's reliability and trustworthiness" [21]. Expressive value, on the other hand, emphasizes positive returns, such as a feeling of being valued and respected [21]. These elements contribute to affective bonds that develop and are sustained between partners in an exchange [21]. Further, in order for a reciprocal exchange to convey symbolic value, three conditions must be met [21]. First, the exchange in question should recur over time. Second, when an act is initiated, there should be no expectation of immediate return. This means that there should be no negotiation, formal agreement, or structures in place to guarantee immediate reciprocity. Finally, in order for a reciprocal exchange to convey symbolic value, it must be a voluntary choice by a recipient to return benefit to the donor. This characteristic would most easily apply in reciprocity for mutual respect rather than reciprocity for mutual benefit.

In summary, while instrumental value of reciprocity aims to enhance the individual utility of the recipient, symbolic value will mainly focus on the social solidarity of a relationship [21].

5.8 Conclusion

In an effort to present a concept that will be an appropriate conceptual basis to relational autonomy when framing the duty to inform of researchers and the disclosure of information to population biobank participants during the consent process, this chapter undertook to introduce and outline the concept of reciprocity. I demonstrated that reciprocity involves an exchange between a donor and a recipient. More importantly, I noted that a reciprocity analysis intrinsically focuses on how recipients respond to donation.

I highlighted two major conceptions of reciprocity: reciprocity for mutual benefit and reciprocity for mutual respect, each with its own set of criteria. In order to later adapt these conceptions to population biobanking, I examined a number of different possible characteristics of the exchange between donors and recipients. These included the nature, scope, flow, and value of each exchange. Of course, the various attributes I have described do not necessarily come in uniform packages: that is, reciprocity for mutual benefit does not necessarily entail that exchanges at its core will be individual in nature, negotiated in flow, nonspecialized in type, or instrumental in value. On the contrary, the various conceptions of reciprocity may have varied reciprocal exchanges, sometimes combining different types or values at once. Understanding these intricacies will help to better adapt the concept of reciprocity in the population biobanking setting, with all of its nuance and associated caveats.

More importantly, in order to demonstrate that reciprocity is an appropriate grounding for relational autonomy—a conception of autonomy that will need to be respected by researchers when disclosing information to participants during the consent process—it was critically important to present the concept of reciprocity as a first step in an analysis that will ultimately require autonomy to be understood through the prism of relationships between stakeholders, rather than through the lens of self-interested individualistic considerations. This will then allow me to characterize the resulting conception of autonomy and how it will affect the disclosure of information by researchers during the consent process.

As I discussed in previous chapters, the individualistic conception of autonomy faces numerous challenges in the context of population studies. In the next Chapter, I will demonstrate that, despite certain limitations, reciprocity is the most appropriate grounding for relational autonomy in the population biobanking context. This is so, in large part, because of its ability to acknowledge and sustain the complex, ongoing and multilateral relationships established by these research projects without also compromising the correlative rights of research participants.

References

[1] Downie J, Sherwin S. A feminist exploration of issues around assisted death. Louis U Public L Rev 1996;15(2):303–27.

[2] Knoppers BM, Chadwick R. Human genetic research: emerging trends in ethics. Nat Rev Genet 2005;6(1):75–6.

[3] Deschênes M, Cardinal G, Knoppers BM, Hudson M, Labuda D, Bouchard G, et al. Énoncé de principes sur la conduite éthique de la recherche en génétique humaine concernant des populations. Montreal: Réseau de médecine génétique appliquée; 2010. https://www.rmga.qc.ca/admin/cms/images/large/encartfran_2609_2e_001.pdf [Accessed 21.03.19].

[4] Beauchamp TL, Childress JF. Principles of biomedical ethics. 6th ed. Oxford: Oxford University Press; 2001.

[5] Gouldner AW. The norm of reciprocity: a preliminary statement. Am Socio Rev 1960: 161–75.

[6] Cohen MB. Perceptions of power in client/worker relationships. Fam Soc 1998;79(4): 433.

[7] Eriksen KÅ, Sundfør B, Karlsson B, Råholm M-B, Arman M. Recognition as a valued human being: perspectives of mental health service users. Nurs Ethics 2012;19(3):357. cited in Sima Sandhu et al. Reciprocity in therapeutic relationships: A conceptual review (2015) 24 International J Mental Health Nursing. p. 464.

[8] Hem MH, Pettersen T. Mature care and nursing in psychiatry: notions regarding reciprocity in asymmetric professional relationships. Health Care Anal 2011;19(1):65–76.

[9] McCann T, Clark E. Advancing self-determination with young adults who have schizophrenia. J Psychiatr Ment Health Nurs 2004;11(1):12–20.

[10] Stephens C, Breheny M, Mansvelt J. Volunteering as reciprocity: beneficial and harmful effects of social policies to encourage contribution in older age. J Aging Stud 2015;33: 22–7.

[11] Johnson TC. Reciprocity as a foundation of financial economics. J Bus Ethics 2015; 131(1):43–67.

[12] Viens A. Public health, ethical behavior and reciprocity. Am J Bioeth 2008;8(5):1.

[13] Kleinman A. The art of medicine care: in search of a health agenda. Lancet 2015; 386(9990):240–1.

[14] Becker LC. Reciprocity. Chicago: The University of Chicago press; 1993.

[15] Becker LC. Reciprocity, justice, and disability. Ethics 2005;116(1):9–39.

[16] Hartley C. Two conceptions of justice as reciprocity. Soc Theor Pract 2014;40(3): 409–32.

[17] Rawls J. Political liberalism. New York: Columbia University Press; 2005.

[18] Hobbs A, Starkbaum J, Gottweis U, Wichmann H, Gottweis H. The privacy-reciprocity connection in biobanking: comparing German with UK strategies. Public Health Genomics 2012;15(5):272–84.

[19] Macneil IR. Exchange revisited: individual utility and social solidarity. Ethics 1986; 96(3):581–93.

[20] Molm LD. The structure of reciprocity. Soc Psychol Q 2010;73(2):119–31.

[21] Molm LD, Schaefer DR, Collett JL. The value of reciprocity. Soc Psychol Q 2007; 70(2):199–217.

Toward a reciprocity-based relational autonomy for population biobanks: advantages and limitations

6.1 Introduction

In this chapter, I demonstrate that, despite certain limitations, reciprocity is the most suitable conceptual grounding for relational autonomy in population biobanks. The result of which, as I will show, is a more appropriate conception of autonomy that is capable of theoretically framing the disclosure of information during the consent process. This is so because, when compared to individualistic autonomy, reciprocity-based relational autonomy offers a more solid basis on which complex, ongoing, and multilateral relationships can both be acknowledged and sustained.

The relationship between the disclosure of information, autonomy, and its relations is crucial, and has been a recurring theme throughout this book. In Chapter 1, I began with an examination of observable characteristics of the duty to inform through the lens of two leading Canadian court decisions: *Halushka* and *Weiss*. There, I showcased how requirements underlying the researcher's duty to inform participants were higher in intensity when compared to the clinician's duty to inform patients. By examining the origins of the exacting duty to inform favored by Canadian courts, I showed how an individualistic conception of autonomy—rooted in liberal individualism—was at the core of the traditional duty to inform. In Chapters 2—4, I demonstrated how population biobanks, which are longitudinal, international, and less directly focused on individuals than conventional research, challenge this conception of autonomy. More specifically, I showed that by adopting a unidirectional focus on the participant, important considerations (including benefit considerations) relating to other stakeholders, namely the public and the research community, end up being overlooked. Indeed, this individualistic focus requires that population biobanks reconsent participants every time a researcher accesses their data and samples. With projects averaging more than 10,000 participants, doing so risks creating delays and impeding the timely sharing of data and samples. Ultimately, this may hamper the return of enriched data emanating from the use of the data and samples by members of the research community, in turn frustrating the orderly translation of knowledge to the clinic, and by extension, to the public at large.

Reciprocity in Population Biobanks. https://doi.org/10.1016/B978-0-323-91286-0.00012-5

At the end of Chapter 4, I examined relational autonomy and considered how it potentially coheres with multilateral relationships, rather than uniquely focusing on individuals. However, in order to practically apply relational autonomy in the population biobanking context, it is first necessary that it is complemented by a concept capable of accurately describing, acknowledging, and sustaining the relationships in population biobanks. As a first step toward that goal, Chapter 5 introduced the concept of reciprocity. I presented reciprocity's key elements: (1) the presence of a donor and recipient and (2) the existence of a reciprocal exchange. I then examined the possible nature of the relevant reciprocal exchanges: their scope, flow, and the values bestowed on them. Doing so was aimed at constructing the theoretical underpinnings necessary to apply the concept of reciprocity as a complement to relational autonomy. In this chapter, I will use these underpinnings to show how the resulting conception could effectively palliate the shortcomings of individual autonomy by accounting for and sustaining the multilateral relationships and interactions that are at the heart of population biobank research projects while at the same time protecting research participants when disclosing information to them during the consent process.

Chapter 6 will begin with an overview of the treatment of reciprocity in biobanking literature (Section 6.2). Given that this book aims to assess the extent to which reciprocity-based relational autonomy—a conception to be respected in the disclosure of information to participants during the consent process—is congruent with population biobanking, understanding how reciprocity has been understood in past biobanking literature will provide relevant and necessary background. This work will help to (1) assess the degree to which reciprocity features in scholarly work and (2) determine whether autonomy has ever been a key consideration when reciprocity has been discussed in the biobanking literature.

This chapter will also draw on the theoretical framework presented in Chapter 5 to properly describe the multiple reciprocal relationships that exist in the population biobanking context. More specifically, in Section 6.3, I will outline three specific reciprocal relationships: that between population biobanks and participants, between population biobanks and the public and between population biobanks and the research community. For each, I will identify the relevant donors and recipients as well as the nature, scope, flow, and value of the possible reciprocal exchanges between them. Crucially, as much as possible, I aim to provide a nuanced and detailed account of how reciprocity can best describe the existing relationships between all of the stakeholders implicated in population biobanking. Using these relationships as a guide, I then demonstrate how they may be understood according to relational autonomy and how such understanding ultimately affects the disclosure of information to participants during the consent process. To that end, I will explain how the new conception of autonomy will be exteriorized when disclosing information to research participants and how this differs from the current approach used in population biobanks (Section 6.4). Finally, I conclude this chapter by examining the advantages and legal limitations of introducing reciprocity as a basis for relational autonomy in population biobanks (Section 6.5).

6.2 The concept of reciprocity as portrayed in biobank literature

This section aims at understanding how the concept of reciprocity has been portrayed in past biobanking literature. Undertaking this literature review, in light of my proposal to situate relational autonomy in the conceptual framework of reciprocity, is important for two reasons. First, it will allow me to be cognizant in my analysis of reciprocity of the various discussions on the topic as a way of ensuring that I do not, so to speak, reinvent the wheel when adapting the concept of reciprocity to population biobanks. Second, and for similar reasons, this literature review will help to determine whether autonomy more specifically has ever been a key consideration in discussions about reciprocity.

A review of the existing literature on reciprocity in the context of biobanking reveals that discussions are limited in a number of important ways. In the following few paragraphs, I aim to highlight how this is so by presenting the different ways reciprocity has featured in the literature. By the end of this review, I will have demonstrated that the literature cannot, at present, form the basis for the working model of reciprocity-based relational autonomy I wish to present in this chapter.

A first general limitation of the literature is that, while most articles engage with the concept of autonomy, very few provide an in-depth analysis of its relationship to reciprocity. The influential Knoppers and Chadwick paper on emerging trends in the ethics of human genetic research [1], for example, describes reciprocity, along with universality, mutuality, citizenry, and solidarity, as novel concepts in research ethics [1]. These concepts are said to embody the complexity of contemporary research endeavors and reflect the growth of public participation [1]. Knoppers and Chadwick describe reciprocity as a form of "recognition of the participation and contribution of the research participant" [1]. Beyond that, they propose broadening the concept of reciprocity as a form of exchange that includes both individuals and the general population [1]. In doing so, they recognize an important role played by actors outside of the participant—researcher relationship. However, the article does not deliberate on the concept of reciprocity in greater detail and does not consider how it would practically be used in relation to autonomy. Furthermore, the research community does not appear to have been captured in the discussion.

The Knoppers and Chadwick article is but one example of an article in which reciprocity is mentioned briefly. Other articles that succinctly mention reciprocity are primarily concerned with public engagement in biobanking research generally [2]. In one such article, Gottweis et al. argue that reciprocity may play a role in addressing important socio-ethical issues raised in biobanking, such as privacy and benefit sharing [2]. While they do not define reciprocity, they claim that "people need to feel that they are part of something larger and that their donation feeds into a mutual, respectful relationship" [2]. On their view, reciprocity facilitates this interaction by creating a "culture of care for the study participants and transparency that is integral to biobank" [2]. Again, while presenting interesting angles from which to view reciprocity—especially those that highlight the importance of mutuality and respect—the Gottweis et al. article does not offer a practical understanding of how the concept of reciprocity could be used to better understand the autonomy of participants when information is being disclosed by researchers during the consent process.

A second limitation revealed in the literature is that examinations of reciprocity in the context of biobanking typically only discuss the relationship between biobanks and participants or biobanks and the public. While these relationships are surely important, utilizing the concept of reciprocity in the interactions of members of the research community—or within a multilateral sphere where all the stakeholders' interactions influence each other—is ignored. Articles discussing the relationships between biobanks and participants or biobanks and the public tend to outline the views and expectations of participants in biobanking research and other kinds of disease-specific projects, including those in which recruited participants are unhealthy (for example, cancer patients). Most of these articles operate on the view that research participants do not usually provide data and samples purely altruistically [3]. This is supported by the proposal that, as some authors have pointed out, biobank participants do not simply forget about the bodily substances they have donated, but rather maintain "a complex relationship with their removed, but not completely detached or disentangled" [4] biological samples. All of the articles reviewed indicate that participants place a great deal of importance on donating their data and samples for research. At the same time, these articles also identify a need for participants to receive something in return for their participation. Interestingly, this expectation is not limited to participants with an illness or condition, where some form of personal therapeutic benefit might easily be anticipated. Healthy volunteers typically have similar expectations, for example, they might understand future familial or social benefit as an extension of *personal* benefit [3]. Participant surveys have shown that most embrace what authors refer to as "reciprocity," whereby participants wish to feel they are taking part in "something larger and that their donation feeds into a mutual, respectful relationship" [5] that features in a complex "social exchange" [6]. In one Australian study, members of the public were asked to complete a survey assessing their beliefs about trust, intention, and benefit implicated in biobank participation [7]. Results indicate that a large majority of participants endorsed reciprocity. In fact, survey participants reported an expectation that personal benefits would be returned to biobank donors [7]. For these participants, such return is an intuitive question of fairness: "reciprocal behavior can be viewed as a desired end in itself as a fair method of distributing resources. Those sharing their resources […] should receive something back […] simply because it is considered the fair thing to do" [7]. In these articles, the discussion around reciprocity is more substantial than in such papers as those authored by Gottweis et al. or Knoppers and Chadwick, all of whom considered the concept at a higher level of generality. However, articles discussing reciprocity as a relationship between biobanks and participants or biobanks and the public tend to limit their presentation on participant expectations, without delving deeper into how a concept such as reciprocity can play a comprehensive role in the way researchers communicate with and inform participants.

The third limitation I encountered in my literature review originates in articles solely focused on the concept of reciprocity and its theoretical underpinnings. In contrast to those reviewed above, this set of articles do not contemplate reciprocity generally, but discuss its use in the field of biobanking. I consider them limited to the extent that they invariably discuss reciprocity in a manner that suggests the creation of an obligation to return and, by extension, that biobank participation should be conditioned on such return [8]. While it may be that some possible conception of reciprocity includes this characteristic of conditionality, it is surely not a necessary feature. Nadja Kanellopoulou, writing from a governance perspective, has contributed substantially to this debate. In one article, Kanellopoulou sets out to address the imbalance of legal power between researchers and participants in the biobanking context [8]. On her view, the notion of research participants conceiving of the donation of data and samples as an unconditional gift is false [8]. Instead, she proposes applying reciprocity in biobank governance in order to encourage engagement, cooperation, and trust between participants and researchers. She writes the following:

> I propose that a better approach for law to protect participants' interests would be to focus on the nature of their [participants'] interaction with researchers and describe it as an ongoing cooperation and dynamic relationship with special obligations for both sides [8].

She sees this approach as a way to empower research participants and encourage a more balanced relationship between researchers and participants [8]. Moreover, the adoption of reciprocity would demonstrate to research participants that the contributions they have made are valued and respected [8]. Against this backdrop, Kanellopoulou proposes that donated samples and data should be considered conditional gifts that extend from biobank participants to researchers. In order for this framework to function, Kanellopoulou suggests that participants and researchers must agree to return conditional gifts to each other in ways that protect participants and does not impede research [8].

A few years later, in a second chapter on reciprocity in biobanking, Kanellopoulou continued promoting this view of reciprocity as an empowerment tool that seeks to restore balance in the relationships of researchers and participants. Her argument cites major biobanking initiatives in the United Kingdom as a ground for refuting the assumption that participants are primarily motivated by altruism. Instead, Kanellopoulou calls for wider participant control exercised through reciprocity [9]. In her view, the return of reciprocal benefit sustains cooperation and trust [9]. Kanellopoulou thus argues that proposals that allow participants to benefit from their contribution should be taken seriously, even if returned benefits are small or intangible. In realizing this goal, Kanellopoulou calls for mutual understanding and agreement between researchers and participants [9]. In her two contributions, Kanellopoulou proposes a number important elements for consideration, including that seeing altruism as the sole reason participants enroll in research is inaccurate. Kanellopoulou also emphasizes the role of reciprocity as an underlying notion that would allow participants to feel valued and respected. The overarching limitation, however, is that she does not discuss reciprocity from the perspective of relationships between biobanks and the public or biobanks and the research community. Instead, her work focuses mainly on reciprocity between the biobank researcher and the participant. Furthermore, the notion that reciprocity only fits within a "conditional" exchange lacks nuance. As I discussed in Chapter 5, certain reciprocal exchanges are seriate in nature, which means that they are not negotiated in advance and are not conditional. Suggesting that the enrollment of biobank participants is conditional on some sort of return that protects participants and does not impede research endeavors [9] is one dimensional. It does not, after all, account for all of the possible exchanges between participants and researchers. This is, admittedly, something Kanellopoulou acknowledges. She concludes her chapter with the prescient observation that "workable notions of reciprocity in the evaluation of participants' contribution in research" [9] is conspicuously absent in contemporary reciprocity literature. For Kanellopoulou, much of that work remained to be developed. In essence, what this literature review informs us is that there is still work to be done toward understanding how reciprocity can reflect the contributions made by participants in the research setting. In a way, such realization is an important premise for what this chapter aims to demonstrate. Basically, to provide a workable notion of reciprocity, I will need to use it as a conceptual framework that practically describes, acknowledges, and sustains the multilateral relationships implicated in population biobanks, which will help me form the basic understanding of the relations embedded in the relational conception of autonomy that I propose to adapt.

More specifically, using the theoretical foundation reviewed in Chapter 5, Section 6.3 of this chapter will attempt to provide a workable notion of reciprocity by examining the nature and characteristics of the possible exchanges between the population biobank researcher and three other stakeholders, namely (1) the participant, (2) the public, and (3) the research community. More importantly, I will also demonstrate how notions of reciprocity between the population biobank and the participant cannot simply be studied in the abstract, but must include tangible considerations emanating from the reciprocal exchanges between the population biobank and other stakeholders. In the section below, I present a conceptual framing of reciprocity that is reflective of the reality of population biobanking and that can serve as a practical grounding for relational autonomy. In this book, I aim to bring together several strands of thinking about reciprocity that are present in different bodies of literature. From these, I propose an approach with sufficient specificity to facilitate new thinking about the disclosure of information to participants in the population biobank setting. My approach differs from the frameworks seen in the literature review above to the extent that it is built on a comprehensive understanding of reciprocity, its attendant conceptual framing, and its more pointed emphasis on the importance of including all relevant stakeholders in the analysis. While the researcher–participant relationship is of interest in assessing the disclosure of information by researchers, other relationships must not be ignored. Indeed, accounting for all of the implicated actors may be the most neglected aspect of discussions surrounding autonomy and reciprocity in the field of biobanking.

6.3 Reciprocity-based relational autonomy for population biobanks or the importance of considering all stakeholders

Reciprocity is generally thought to be associated with such elements as trust, respect, and mutuality [8,10]. It is unclear, however, how these components could be meant to work together in the context of population biobanking. Is trust, for example, a necessary condition for reciprocity or merely its consequence? Does mutuality have meaning beyond a simple mutuality of benefit? Will all possible exchanges in the context of population biobanks be dependent on some form of reciprocation? What is the role of "respect" or the recognition of one's contribution in this framework? I will attempt to answer these questions in the following sections, keeping a view toward understanding how reciprocity can provide a plausible basis for the concept of relational autonomy. When I examined the current jurisprudential interpretation of the duty to inform of researchers and the individualistic conception of autonomy that is at its core, I analyzed the type of relations that existed between the actors found within that framework, namely, participants and researchers. Now, I will demonstrate how reciprocity can form a suitable grounding for relational autonomy in considerations surrounding the disclosure of information to participants during the consent process. To do so, I need to understand the "relations" that are described by the relational conception. This is why I will first examine the possible exchanges that are at the heart of this reciprocal relationship. Once that is complete, I will be able to better qualify the resulting conception of autonomy and highlight how it will affect the way in which population biobank researchers will satisfy their duty to disclose information to participants as part of the consent process. I will not limit my examination of the relevant exchanges to those between the population biobank and the participant. Doing so would imply that the exchanges between these two stakeholders can be studied independently of any other consideration, which would not truly differ from the approach taken in the individualistic conception of autonomy. I will still examine exchanges between the participant and the population biobank, but only after having studied exchanges between the population biobank and the public (Section 6.3.1) and the population biobank and the research community (Section 6.3.2). From there, I will demonstrate both that all of these stakeholders are part of multilateral reciprocal relationships (see Fig. 6.1) and that notions of reciprocity between the population biobank and the participant (Section 6.3.3) must include considerations emanating from reciprocal exchanges between the population biobank and the other stakeholders. In order to provide a workable notion of reciprocity in the multilateral relationships found in the population biobank context, I will refer to theoretical notions seen in Chapter 5 and apply them as necessary in this chapter. For example, I will highlight the nature, scope, flow, and value of each reciprocal exchange under study.

As a note, I do not think that population biobanks can themselves be plausibly considered moral agents. For reasons of brevity, I only use the term "population biobanks" to refer to researchers overseeing population biobanks.

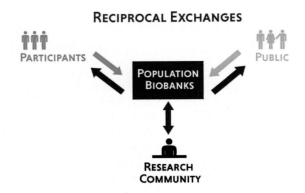

FIGURE 6.1

Reciprocal exchanges in population biobanks.

6.3.1 The population biobank—public relationship

As seen in the literature review above, discussions of reciprocity in the biobanking context generally focus on implications for participants and the public. Considering the history of individual-centered discussions, this shifting interest from individuals to the broader population [1] is a positive development. This section will dissect the population biobank—public reciprocal relationship and provide granularity absent in the current literature. This section will begin by studying the reciprocal exchanges possible in cases where the public is the donor (Section 6.3.1.1) before shifting to a similar analysis when the public is a recipient (Section 6.3.1.2). Looking at both possibilities will allow a comprehensive understanding of this reciprocal relationship.

6.3.1.1 When the public is a donor

In the relationship between population biobanks and the public, I argue that the public primarily plays the role of the donor. Indeed, the most prominent source of funding of population biobanks is public money, and by extension, is derived from members of the public at large. This is especially true in the case of Canadian population biobanks [11]. Where this is the case, the population biobank will qualify as a recipient. These roles could conceivably be reversed—a situation to which I will return later.

   Operating on the view that the public is a donor and the biobank a recipient, what type of reciprocal exchanges may be envisaged? First, there must be an act of donation (by the donor—in this context, the public). Secondly, for a relationship to be considered reciprocal, a return by the recipient to the donor must be concluded. As for the act of donation—by the public in this case—it will be the contribution made by members of the public as a collectivity through public funds to create and sustain population biobanks (through tax revenue, for example). Of course, it is certainly true that some biobanks may, in the future, not rely on public funds.

In situations where the population biobank is created and sustained exclusively through private funding, the public may not qualify as a donor, unless members of the public financially contribute to the population biobank as a collective through some means other than taxes. At present, Canadian population biobanks are largely supported by public funds [11]. For this reason, I will focus on this form of donation.

Now that I have established the act of donation, what will the recipient (in this case the population biobank) return back to the public at large? This question is especially important as reciprocity is a concept that focuses on the actions of the recipient following a donation. In other words, we may more precisely qualify the reciprocal relationship in view, thanks to the kinds of return undertaken by the population biobank. In this case, I argue that the conception of reciprocity at the heart of the relationship between the population biobank and the participant is that of reciprocity for mutual respect (see Fig. 6.2). As I outlined earlier in Chapter 5, the foundational view of the mutual respect conception of reciprocity is that its ultimate purpose does not turn on sustaining a mutually advantageous relationship, but rather to extend thanks to the other party, to show respect by recognizing the other's contribution [12]. More specifically, I posit that there are three possible mechanisms for the population biobank to reciprocate to the public. I also believe that all three will respect conditions of fittingness (being a good, being seen as a good, and being seen as a return by the donor) and sufficiency (aim to show respect and acknowledge the contribution made by the donor).

The first kind of return by the population biobank is centered on the implementation of efficient access mechanisms to data and samples stored by the biobank. The implementation of access mechanisms not only involves the development of documentation necessary to support them, but the creation of bodies tasked with evaluating and approving access requests as well [13]. Such return actually engages two other stakeholders: the participant and the research community. In fact, the goal of implementing efficient access mechanisms is to provide the research community with the ability to access data and samples of participants to further their own research projects and enrich the population biobank. This is a short-term goal. The long-term goal is that through access and enrichment, new discoveries will be made possible, which may in turn benefit the population as a whole. Indeed, the primary goal is to increase the statistical power needed to generate useful results, which, in turn, will translate into meaningful knowledge [14] for society [15] and future generations. The ultimate purpose is to improve the health of the population and correlatively increase public trust once such outcomes are materialized [16].

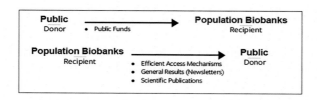

FIGURE 6.2

Public—population biobanks reciprocal relationship.

Implementing efficient access mechanisms is both fitting and sufficient. It is fitting to the extent that it represents a good and is likely to be seen both as a good and a return by the general public [17]. The public as a collectivity might not be aware of all the inner workings of a population biobank when it provides public funding. But that does not mean that the population biobank should not need to ensure that such funding is well utilized. Implementing efficient access mechanisms is one way of ensuring that funds given by the collectivity are properly utilized and that knowledge emanating from data and samples is maximized. Implementing efficient access mechanisms is also sufficient in the sense that its goal is not the balancing of benefit with the donor, but rather of showing respect to them as contributors and recognizing their contribution. The population biobank will do so by ensuring that proper access mechanisms are in place that allow the participant's donation to be used efficiently and in accordance with what they were promised during the consent process. This is illustrated in Table 4.1 of Chapter 4, above. The act of return also aims to contribute to a cooperative project, seeking to promote the general health of the population. It will be difficult to see how such return could not be justified to all members of society, an important indicator of the fulfillment of the criteria of sufficiency.

With this return by the population biobank, the first type of reciprocal exchange is complete. If I were to characterize this exchange, it would be *communal* in nature as it pertains to the public rather than to a particular individual [4]. I also argue that this reciprocal exchange is *generalized*. Generalized exchange is oriented toward maintaining social solidarity and is on the higher end of the spectrum of solidarity-building varieties of reciprocity [18]. When the public donates (in our case, through public funds), they are not entrenched in a relationship that requires commensurable return, which is the opposite of a generalized exchange. Moreover, the flow of the reciprocal exchange can be qualified as *seriate* as it does not feature any agreement or transaction. As for the value of the reciprocal exchange itself, it is symbolic (or more precisely "expressive" [19]). It is not instrumental as it is not negotiated. The act of return by the population biobank to the public aims at letting the public know that their contribution is valued and respected by ensuring efficient access mechanisms to maximize their use [19].

The second kind of return made by the population biobank to the public is through the disclosure of general results emanating from the use of data and samples. Such results are aggregate in nature so as to ensure they do not pertain to specific individuals within the biobank. General results may take various forms, such as newsletters or information made available online. Newsletters, for example, are used prominently by CARTaGENE, which states clearly in its consent form that a yearly bulletin describing research projects that use its resources will be made publicly available [20]. The same process also exists in other population biobanks, for example, in Atlantic PATH [21] and the Alberta Tomorrow Project [22]. Bulletins or newsletters are not the only form of general results disclosure. Today, population biobanks can also publish information and statistics on social media platforms, such as on Twitter. CARTaGENE, for example, posted on its Twitter account that "Almost

25% of CARTaGENE's participants declared having at least one member of their immediate biological family affected by diabetes. This highlights the importance of genes discovery in creating new treatments" [23]. This is an example of an aggregate result. It does not identify specific participants and presents a useful, though general, result based on the assessment of all CARTaGENE participants. These results represent a form of transparency toward the public. By returning such results, the biobank lets taxpayers know that their funds are indeed leading to important discoveries. This interaction or reciprocal exchange falls within a framework of reciprocity for mutual respect. Much like the first type of return, this is both fitting and sufficient. It is fitting because returning general results is a good that contributes to the general knowledge possessed by society. Further, it would be realistic to assume that it will be seen by the public as both a good and as a return. Beyond that, it is also sufficient, as it aims to contribute to members of society in a way that would be reasonable to assume would be accepted by the public. In fact, making such information publicly available sends a strong message to the public and to researchers that the population biobank is producing results. Furthermore, it also aims at educating members of the public about the value of their donation. It is a sensible act of recognition and respect. As for the characteristics of this reciprocal exchange, it is *communal* in nature, *generalized* in scope, *seriate* in terms of its flow, and *expressive* as to its value for the same reasons as those presented for the first kind of return above (i.e., implementing efficient access mechanisms).

A third type of return concerns the dissemination of information in scientific conferences and through publications based on discoveries that use data and samples from the biobank. Researchers in population biobanks disseminate and publish articles using data and samples as a way to contribute to the scientific literature [24]. While journal articles are generally not intended for broad public consumption, they are nevertheless helpful for advancing our current thinking and fostering an environment of collaboration and innovation that can be expected to materialize in downstream scientific applications of broader benefit to society. More importantly, some discoveries from scientific publications are presented to the public through mainstream media. News stories such as "For Women, Confusion About Alcohol and Health" published in the *New York Times* are a good case in point [25]. In this particular case, the medium in question—although not Canadian—cited a study conducted by a population biobank in the United States that showed that the number of alcoholic beverages consumed per day, not type of beverage, is associated with an increased risk of developing breast cancer in women. At the time, the study in question was presented at a European conference [26] and was later the subject of a scientific publication in the *European Journal of Cancer* [27]. Dissemination of discoveries emanating from this study informed the general public. Furthermore, population biobanks funded through public funds will acknowledge such funding in any publications. In their marker paper, researchers with CARTaGENE thanked Genome Canada and Genome Quebec, the primary funders of their cohort [28]. Both Genome Canada and Genome Quebec receive government funding through budgets adopted in the federal and provincial legislatures.

In my view, the reciprocal exchange described above exhibits reciprocity for mutual respect. The return made by the population biobank is fitting; it is a "good" and can reasonably be considered by the public to be both a good and a return. Further, it aims at contributing to a social project. In fact, the population biobank is showing respect and recognition by striving to inform the public about their health and the risks associated with different types of consumption. It also showcases to the participant how their donation has now bore fruit by generating new research findings that are presented to the public. Population biobanks are under no particular obligation to do so, but by reciprocating in this way, they send a message to the participant that their original donation was valued and accordingly efficiently utilized.

For exactly the same reasons as the other two kinds of returns listed in this section, I contend that the reciprocal exchange created following the dissemination of scientific information is *communal* in nature. It is also *seriate* [29] in its flow and *generalized* in its scope. As for its value, it is *expressive* rather than *instrumental*. Indeed, the dissemination of scientific information is not borne out of negotiation, but is a way for the population biobank to emphasize the importance of maximizing on the resources it has been provided through public funds in ways that will ultimately materialize in downstream scientific applications of general benefit to society.

In brief, in the reciprocal relationship between the public as a donor and the population biobank as a recipient, three types of return are possible. All three fit within the framework of reciprocity for mutual respect. Below, I want to briefly examine the possibility of conceiving of the biobank as a donor and the public as a recipient. Following this, I will transition to a discussion of the population biobank—research community reciprocal relationship.

6.3.1.2 When the public is a recipient

The reciprocal relationship between the public and population biobank where the public is a recipient is not as tangible as that in which the public is a donor. For reasons I will lay out below, I would even say that it is difficult to fully conceive. However, I will nonetheless briefly present its rationale in the pursuit of comprehensiveness.

The idea that the public can be a recipient originates from a qualitative study of cancer patient perceptions of biobank research [6], in which authors found that some patients view biobank participation as a way of giving back to science (see Fig. 6.3):

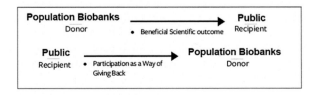

FIGURE 6.3

Population biobanks—public reciprocal relationship.

Patients struggling to face the life-threatening disease and the treatment involved may indeed regard the issue of research as a priority. Whether this is because close relatives have benefited from research, because of moral or civic reasons, or because donation is experienced as a commitment to giving back what one has received, supporting research via biobanking is mainly perceived as a means of promoting and sustaining hope and trust in the future [6].

What this statement presumes is that, among the many factors that could prompt members of the public to participate in research, at least one is the value they collectively receive from discoveries that are made possible by the research undertaken by the population biobank. While it is surely valid to think that population biobanks may potentially work to improve the health of the general public, I think it remains, for the time being, circumstantial and difficult to guarantee in all cases. In fact, it is important to keep in mind that population biobanks are relatively novel undertakings. While they can hold much promise, the fruits of their scientific discoveries will be slow to materialize. This is so, in part, because a number of biobanking projects have 20 or more years remaining before they are expected to reach completion. The promise of value to the public exists, but will require several years to become truly tangible. With that said, we may wonder why this is an issue. It is an issue to the extent that the act of donation must be tangible if it is to initiate a reciprocal exchange. The act of reciprocation, however, can occur at a later time. In other words, when conceiving of the population biobank as a donor and the public as a recipient, the act of donation (in this case, the added value to the public) needs to be tangible and definitive in order for the act of reciprocation to occur (even at a later time). However, because the act of donation is not guaranteed, it is difficult to see how a reciprocal relationship between the population biobank and the public, where the latter is the recipient, can actually exist. The other limitation is that, once members of the public decide to reciprocate (provided—for the sake of argument—that the act of donation is definitive), it would be difficult to conceive of the public as reciprocating through participation. The public here will be substituted by individual participants rendering the reciprocal exchange with the original parties (i.e., the public and the population biobank) difficult to conceive. For these reasons, I will not delve deeper into the matter. I have presented it here simply as a way of presenting all relevant possibilities when studying the reciprocal relationship between population biobanks and the public.

Despite difficulty in conceiving of the public as a recipient, my examination of the reciprocal relationship between the population biobank and the public has established a plausible reciprocal exchange in which the public is a donor and the population biobank is a recipient. This reciprocal relationship is primarily premised on respect and on valuing the donation given by the research participants.

Establishing the existence of such a relationship is crucial. One of my criticisms of liberal individualism as a basis for autonomy focused on the fact that it fails to acknowledge multilateral relationships that are necessarily implicated in population biobank research. Reciprocity, however, does exactly the opposite insofar as it both acknowledges and sustains multilateral relationships that involve the public and other stakeholders. More importantly, in my examination of the reciprocal relationship between the population biobank and the public, I mentioned how other stakeholders might interact within such a relationship. The data and samples, from which scientific discoveries may be derived for the benefit of the public, are donated by participants. When the population biobank implements an efficient access system, it does so while mindful that members of the research community will be accessing the population biobank's resources. The following section will focus on the reciprocal relationship that exists between the population biobank and the research community. This time, the reciprocal relationship will be built on mutual benefit.

6.3.2 The population biobank—research community relationship

Chapter 4 of this book demonstrated that, in order for a population biobank to achieve the statistical significance necessary for the investigation of gene—gene, gene—disease, and gene—environment interactions over time, large amounts of data and samples are required [30]. More importantly, to achieve the requisite breadth, international, regional, and Canadian documents highlighted the importance of collaboration between members of the research community in order to maximize on the use of the data and samples for the benefit of society and future generations.

This section will examine the reciprocal relationship between the population biobank and the research community, a topic long neglected in the reciprocity literature. For this reason, it warrants close consideration of its nature and characteristics. In the population biobank—research community relationship, I posit that the population biobank should be considered a donor and the research community (represented by an applicant requesting access to data and samples) should be considered the recipient. More specifically, the resulting exchange begins when population biobanks provide a member of the research community (research applicant) with access to data and samples (donation).

The aim of this reciprocal relationship is mutual benefit, rather than respect. Indeed, the research applicant requesting access to data and samples is primarily interested in the scientific value of the information these materials contain. The population biobank, correlatively, is interested in the dissemination of data and samples in order to maximize their use, financially sustaining itself through cost-recovery fees paid by research applicants [24], and enriching its dataset from the return of derived data from research applicants after the research project is completed [31]. In this kind of relationship, both entities seek and receive a benefit. Typically, Canadian population biobanks will enter into a formal agreement (known as an Access Agreement [32]) that formalizes the relationship.

In the agreement reached by population biobanks, researchers, and the researchers' institutions, the biobank will commit to providing data or samples for use in a stipulated research project on the condition that governance bodies responsible for adjudicating access requests, such as an access committee, will first evaluate and accept the access request [33]. In return, the researcher applicant (who then becomes an approved user) will be asked to do four things (which I consider returns in the reciprocal relationship and which will be outlined below).

In contrast to reciprocity for mutual respect, reciprocity for mutual benefit must satisfy not only the criteria of fittingness (the return both being a good and being seen as such by the donor), but also the criteria of proportionality (rather than sufficiency). Given that the relationship between population biobanks and researchers is modulated by an agreement in which the biobank and researcher have equal bargaining power, both parties will agree only to an exchange of what they both consider qualifying as a good. As a result, this relationship will necessarily satisfy the criteria of fittingness. Proportionality, in turn, requires that the relevant exchange be balanced. As I demonstrated in Chapter 5, such balance can be understood as either a return of commensurate benefit with as little sacrifice as possible, or an equal marginal sacrifice in which the return is proportional to donor sacrifice [34]. Depending on the nature of the return made by the research applicant, I believe they can either fulfill the criteria of proportionality based on commensurate benefit or based on equal marginal sacrifice. This section will review four types of return and establish which understanding of proportionality they seem to fulfill.

First, the population biobank will require that applicants pay a cost-recovery access fee prior to receiving the data or samples. Such fees are used to defray administrative and operational costs associated with making data and samples available [35]. These exchanges are proportional in nature as they are premised on an equal sacrifice in the sense that fees seek only to recover costs and not to earn a profit. The amount of time and work put into preparing and shipping data and samples will be covered by fees set out by the parties. Second, approved users will also be required to maintain strict security safeguards throughout the use of data and samples in order to ensure that the reidentification of participants or unauthorized data and sample access is prevented [36]. This is also proportional as the population biobank will require that the approved user apply similar security safeguards when storing and shipping the data and samples. This is an equal marginal sacrifice. Thirdly, approved users will be required to return data derived from their projects in order to enrich the population biobank's database [31]. Such derived data are the culmination of an approved users' research, and sending a copy to the population biobank ensures that data made available to the research community are as updated as possible. In turn, this allows for more efficient research that builds on the work of others [31]. This return can also be viewed as proportional in the sense that it exhibits commensurate benefit with minimal sacrifice. Indeed, the population biobank provided data and samples that were beneficial to the approved user. The latter will return enriched data that will be useful for the population biobank and the research community at large. Finally, the approved user will be required to provide proper attribution to the biobank and its scientific directors (or lead principal investigators) [37] Depending on the level of contribution made by researchers in the biobank, this may lead to a coauthorship in alignment with international authorship standards [38] (see Fig. 6.4). This again signifies commensurate benefit with minimal sacrifice.

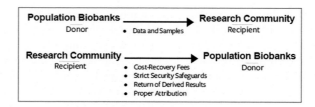

FIGURE 6.4

Population biobanks—research community reciprocal relationship.

Accessing data and samples will allow approved users to further their own research and to publish. Proper attribution will also benefit the biobank researcher who may have his/her name featured on the same paper depending on their level of contribution. This would be commensurate benefits with minimal sacrifice.

As far as the nature of the reciprocal exchanges mentioned above is concerned, I would suggest that they are individual in nature. Even though the term "individual" usually concerns one particular person, I would still maintain that it is individual in the sense that it concerns only one party (the population biobank or its researchers). It cannot be seen as communal because it does not pertain to a larger group of people as a group per se. As for the scope of the reciprocal exchanges above, I believe they are nonspecialized as they occur within an agreed upon timeline and include commensurate goods [18]. The flow of the reciprocal exchange would be *negotiated* given that returns made by the member of the research community would be based on a prior agreement that specifies what each party will receive from the other [29]. This is especially true of access agreements signed between population biobank and approved users. Not only do such agreements contain clear terms and conditions, but they will also lay out the responsibilities of the parties (and the nature of what each party will do and return) and contain a jointly acceptable timeframe for fulfillment of the agreement [39].

In contrast to the relationship between population biobanks and the public (which holds an expressive value), the value underpinning the relationship between population biobanks and members of the research community is instrumental [19]. Instrumental value, also known as utilitarian value, refers to acts of reciprocity that provide some form of utility to the recipient: "their value is instrumental in the sense that they help the recipient meet the need that was the original objective of the exchange" [19]. This is exactly what access agreements, which are a foundational element of the relationship between the biobank and members of the research community, aim to achieve. Each party will receive jointly negotiated benefits that will help them to accomplish their unique objectives.

The previous subsections aimed at presenting two kinds of reciprocal relationships: the first was between population biobanks and the public, while the second was between the population biobank and the research community. During this examination, it has been clear that these relationships—although mainly specific to the parties involved in them—all implicate the biobank participant in some way. For example, in the population biobank—public relationship, one of the mechanisms of return to the public included the implementation of an efficient access system for the use of data and samples donated by participants. In the population biobank—research community relationship, one of the mechanisms of return by the population biobank to the participant included ensuring that strict security safeguards that protect the data and samples of participants are put in place. These examples show that it will be difficult to look at the population biobank—participant relationship without accounting for relations involving the public and research community. Indeed, only the concept of reciprocity has the ability to acknowledge and sustain these multilateral relationships between different stakeholders. The section below will specifically examine the reciprocal exchanges between the population biobank and participants and highlight, where possible, how they interact with both the public and research community.

6.3.3 The population biobank—participant relationship

In this section, I examine the reciprocal relationship that exists between population biobanks and research participants; a reciprocal relationship that I will demonstrate is rooted in respect rather than mutual benefit. I will also demonstrate that notions of reciprocity between these two parties include considerations emanating from the reciprocal exchanges between the population biobank and the other stakeholders studied in the last sections, namely the public and research community.

The existence of a donor—recipient relationship is, as we saw in Chapter 5, the first requirement in a relationship of reciprocity. It should not be controversial to propose that the donor in this case is the participant who donates data and samples to the population biobank. Correspondingly, it is the population biobank that qualifies as the recipient. A second requirement for reciprocity is the existence of an exchange resulting from a return made by the recipient. I will describe in detail the different kinds of possible return, describing them as either (1) manifest returns or (2) abstruse returns. I have created this nomenclature to differentiate between returns that are tangible and directly affecting participants (manifest returns) and those that are more personal in nature (abstruse returns). For each return, I will describe how the conception of reciprocity at play is one of respect and examine the nature, scope, flow, and value of the resulting reciprocal exchange. Doing so allows me to both identify and describe the type of relation that exists between the population biobank and the participant in order to practically apply these terms in descriptions of the relational autonomy at play and how it affects the disclosure of information by researchers in population biobanks.

6.3.3.1 Manifest returns
In this subsection, I will explore what I call manifest returns in order to better understand the different facets of the participant—population biobank relationship. There are four kinds of returns under this category. More specifically, in return for their donation, the population biobank offers participants the following: a promise to

safeguard their privacy beyond the storage period, communicate with them in an ongoing fashion, return abnormal results during the assessment center visit, and return individual research results (IRRs) and incidental findings (IFs) where possible.

The first way in which population biobanks reciprocate to participants is through the establishment of safeguards that protect the confidentiality of stored data and samples beyond storage [40]. The population biobank will do this primarily during the collection and storage of data and samples, but will continue doing so when it shares such samples with authorized members of the research community. Procedures are put in place not to guarantee *full* confidentiality, but rather to make reasonable efforts to limit the possibility of unauthorized access to the data and samples [40,41] This is an effort that requires both good storage practices when the data and samples are located within the population biobank and strong confidentiality rules when the data and samples are accessed by the research community. For external access by researchers, an efficient access governance system must be established in order to adjudicate requests, applicant credentials, and operational readiness. Some population biobanks go so far as to request that approved users provide copies of articles before they are submitted for publication in order to ensure that participants have not been identified [37]. I consider the resulting exchange to be indicative of reciprocity of mutual respect because the return strikes me as both fitting and sufficient. It is fitting as what is being returned by the recipient (i.e., safeguards to protect confidentiality beyond the storage period) is a good for the donor (participant) and can reasonably be considered as both a good and as a return. It is also sufficient if it aims at acknowledging the donor's contribution in a fair way and showing proper respect for it. A skeptic might assert that protecting the confidentiality of the data and samples or participants is a legal requirement and establishing safeguards to protect it is an obligation imposed by law. While this is true, the extent to which population biobanks go to protect it, however, is indicative of a willingness to value a participant's contribution (even when data or samples are no longer under their control). During collection and storage, the population biobank is bound by the general requirements of confidentiality. It may also elect to place the onus on an approved user to ensure that confidentiality is protected. However, most population biobanks have enacted safeguards that go beyond the storage period and extend to overseeing the use of the data and samples when they are no longer under their control. Indeed, population biobanks, such as the Canadian Partnership for Tomorrow Project, a consortium of five major population biobanks in Canada, include an auditing clause in their Access Agreement allowing them to assess whether the host institution currently using the data and samples properly protect them [42].

When the recipient protects the data and samples of donors beyond the storage period and when they are no longer under their control, the recipient population biobank is not aiming to produce a balanced exchange of benefits, but is rather acknowledging the importance of the donation and is doing everything possible to protect it. The resulting reciprocal exchange can be identified as individual in the sense that it concerns the research participant in question. Beyond that, the scope of the reciprocal exchange is generalized. When the participant donates their data and samples, they are not bound by a relationship that requires commensurable return. Moreover, the flow of the reciprocal exchange can be qualified as *seriate* as it does not feature any agreement or transaction. Finally, the value of the reciprocal exchange itself, is "expressive" [19]. Indeed, putting in place proper safeguards to protect a participant's privacy aims at letting them know that their contribution is valued and, ultimately, respected [19].

The second type of reciprocal return made by population biobanks to participants takes the form of continued ongoing communication. In order for the biobank to keep participants informed about its activities, it maintains regular contact with participants. Such contact will offer an opportunity to ask participants new questions, invite them to participate in new projects or request that they provide new samples [20]. Above all, such communication is an opportunity to inform participants about study progress and outcomes, usually through the publication of regular newsletters [21,22]. Another example of ongoing communication may be found in public registries. The Canadian Partnership for Tomorrow Project [43], for example, has created a website containing a public, openly accessible registry of ongoing research projects that are currently using samples and data [44]. The goal is to allow research participants—as well as the general public—to learn more about an approved user, their affiliation, and to access the lay summary of their project. Much like in the first kind of return, ongoing communication also falls under reciprocity for mutual respect. It is fitting as ongoing communication is a good and research participants can reasonably consider it both a good and a return. It is also sufficient insofar as it acknowledges the donor's contribution. The ultimate aim is to thank participants and show them respect. Indeed, by highlighting new discoveries or recent publications only made possible through the use of their data and samples, the population biobank is interested to show participants the extent of the value of their donation and concretely indicate that their data and samples have worked to advance scientific knowledge. This information can be made available to participants, but this would require that they either request them personally or they search for mentions of the relevant biobank in scholarly publications. Through proactive ongoing communication, population biobanks aim to save participants time and effort by making all this information available to them at no cost. The resulting reciprocal exchange is individual in nature, generalized in scope, seriate in its flow, and expressive in its value for exactly the same reasons identified in the first return above (i.e., safeguards to protect privacy).

A third way in which population biobanks may reciprocate is through the return of abnormal findings emanating from physical measurements at an assessment center. Participants who enroll in population biobanks are asked at the beginning of recruitment to visit an assessment center to complete questionnaires and provide samples. During that visit, a number of physical measurements are taken, including bone density and blood pressure. All biobanks return any critical values (those that could pose a serious danger to their lives) discovered during these measurements to participants [45]. Some biobanks could even extend the return period to include abnormal findings from laboratory tests performed before the samples are stored

[45]. In both cases, if critical medical values are identified, the participant is informed. If the situation warrants, they may be escorted to an emergency medical treatment center [45]. Here again, such return falls within reciprocity for mutual respect. Giving back critical health values (such as high blood pressure) to research participants is a good and should reasonably be seen by the participant as both a good and as a return. It is consequently fitting. This third type of return is also sufficient. It acknowledges the donation of the participant and directly contributes to his or her well-being. Providing critical health values respects these participants and their health. Just as for the two other types of return mentioned above, and for exactly the same reasons, the resulting reciprocal exchange from the return of critical health values to participants can only be seen as individual in nature, generalized in scope, seriate in flow, and expressive in value.

Providing critical health values back to researchers raises the issue of a fourth type of return that which involves IRRs and IFs. Such return happens after samples are stored (see Fig. 6.5 below). While IRRs are pertinent to the objectives of a research project (in our case a population biobank), IFs fall beyond its scope [46]. Return of IRRs and IFs should also be differentiated from the return of general research results. While IRRs and IFs concern individual participants, general results concern a group of persons [47]. Both the modalities and conditions for their return differ [45]. I have examined the return of general results in my study of the relationship between population biobanks and the public. Here, I will instead focus on the issue of returning IRRs and IFs to population biobank participants.

Papers discussing the return of IRRs and IFs to participants generally refer to reciprocity as a guiding principle. Some authors have noted that "the obligation to offer results increases when the interaction with a research participant is more extensive because a more intense relationship creates a stronger requirement for reciprocity." [48]. The relationship will be defined in terms of the level of involvement of participants and researchers, the duration of their interaction, and the characteristics of the participant contribution [48]. Put another way, a greater

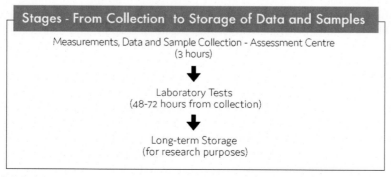

FIGURE 6.5

Storage of data and samples in population biobanks.

degree of participation will entail a greater expectation of return. Bredenoord et al. similarly pointed to reciprocity as a justification for disclosure of IRRs and IFs [49]. Here again, the authors maintain that individuals do not participate in biobank research out of pure altruism [49]. They expect that a productive relationship between themselves and researchers will arise and that such productivity will most tangibly be realized in the form of clinically useful results [49]. This being said, not all authors agree about the place of reciprocity in the debate. Solberg and Steinbekk, for example, authored an article in 2012 on managing the return of IFs and research results in biobanks [50]. While they agree that conceptions of reciprocity and justice play an essential role in structuring relationships between participants and researchers, they contend that any potential value derived from population biobanks makes sense only as it applies to future generations. On their view, beneficence on the part of participants can be understood only at the level of the collective and are never at the level of the individual alone [50].

With that said, an increasing number of authors suggest that IRRs and IFs should be returned to individual participants only when specific criteria are met. The proposed criteria are typically the following: analytical validity, clinical significance, and actionability [51]. In such cases, "analytical validity" refers to the ability to precisely and reliably identify a particular genetic characteristic [51,52], while "clinically significant" and "actionable" findings are those that have a well-recognized and significant risk and for which an accepted therapeutic or preventive intervention is available [51]. In Canada, the *TCPS 2* requires that researchers return material IFs to participants [53]. While prescriptive, this is a vague obligation, for it is unclear what the language of "significance" is meant to convey and how it ought to be assessed. In addressing this concern (and others), the Panel on Research Ethics—a body charged with developing the *TCPS 2*—released a guidance document in 2019 on addressing material IFs. Among other things, this guidance identifies the population biobank setting as one instance in which the circumstances may justify an exception to the general obligation to disclose material IFs, particularly when research participants consent to a policy that does not permit the return of results or when return is impracticable [53]. Return is impracticable, for example, if it cannot be put into practice without jeopardizing the overall conduct of a particular research project [54]. For example, if there are 100,000 participants, systematically returning IRRs and IFs for all participants would require large financial and human resources. If the project is unable to obtain such resources, the imposition of an obligation to return IRRs and IFs risks jeopardizing its overall conduct.

The tension surrounding the return of IRRs and IFs was never higher than in debates surrounding population biobanks. Historically, population biobanks have elected not to return IRRs and IFs after long-term storage. The Public Population in Genomics and Society (P^{3}G) international consortium upheld the validity of this no-return approach [55]. However, given increasing pressure to return validated, significant, and actionable findings, P^{3}G has also proposed that population biobanks may introduce an option of return upon recontact with participants [55]. Recontact is a systematic procedure conducted by population biobanks in which the biobank periodically follows up with research participants while samples are in storage to collect new data or samples or to inform them about ancillary studies in which they may wish to participate. I discussed this in greater detail in Chapter 3, Section 3.4. Alternatively, the P^{3}G *Statement* proposes that the option for return may be inserted in any new consent during the recruitment stage [55]. If the setting is propitious and the participant has consented to receiving IRRs and IFs, then such return may be considered a feature of a reciprocity-based relationship.

In fact, seeing the return of IRRs and IFs materialize in the population biobank setting is not farfetched. While most population biobanks have a no-return policy, some have begun introducing the consent form option in their most recent recruitment processes. The consent form of the Ontario Health Study (OHS) is a good example. For several years, OHS has either been silent or has adopted a no-return policy. Recently, it more directly opened the door to the return of incidental findings:

> *I understand that researchers who use my information and samples in the future might discover something unexpected that could significantly affect my health (known as an "incidental research finding"). [...] I accept that if my information or samples are included in future research, the only time my individual results will be communicated to me is if incidental research findings are found. Otherwise, I accept that the results of future uses of my information and samples will not be shared with me [56].*

Furthermore, biobank participants are likely to be invited to join other projects as part of the recontact process, and some of these projects may utilize new tools that would make the return of IRRs and IFs more realistic. The possibility of joining new projects is most notably available in the case of the Canadian Healthy Hearts and Minds project, in which recruited participants from Canadian population biobanks were sought to perform and store data from MRI scans [57]. Participants who consent to participate in this project are informed that, with their consent, the results of any severe structural abnormality found on the scans would be returned [58,59]. In summary, the return of IRRs and IFs, while subject to much debate, should not be dismissed. The possibility of this happening in the population biobank context has never been more present than with recontact made by new projects of already enrolled population biobank participants. When the criteria of analytical validity, clinical significance, and actionability are satisfied, and when the return from the population biobank to the participant is completed, I believe it becomes another example of manifest reciprocal exchange.

Much like the return of critical health values at the assessment stage, the return of analytically valid, clinically significant, and actionable IRRs and IFs to research participants falls within reciprocity for mutual respect because, in my view, it is both fitting and sufficient. It is fitting in the sense that undertaking such return is a good and can be seen by the participant who has consented to such return as both a good and a return. It is sufficient because it acknowledges the donation of participants by valuing their health and well-being. Here again, the goal is not to create a purely advantageous relationship of benefits, but to acknowledge the donor's contributions and reciprocate in a way that highlights the level of respect they are owed. Population biobanks can always decide not to return any findings to participants. Indeed, for many years, population biobanks have considered themselves as research endeavors not equipped to deal with clinical issues that could emanate from research findings. Hence, returning findings to participants would be considered to be the taking of an extra step. Ultimately, this would aim to send a message to participants that their contribution is valued so much that they are ready to validate and return potentially useful findings in spite of the limitations.

The resulting reciprocal exchange is individual in nature, generalized in scope, seriate in its flow, and symbolic in its value for exactly the same reasons presented in all the above-mentioned returns. One may criticize the fact that the flow is not negotiated in the case of the return of IRRs and IFs given that consent forms will generally simply describe such return and only provide the participant the option of receiving them or not. In other words, it may be said that consent forms represent a kind of negotiation (in reference to a negotiated flow of the reciprocal exchange). This view is incorrect. Consent forms are a snapshot of what was raised and discussed prior to enrollment, but consent is a *continuous process* [60]. It should not be understood to be analogous to the contracts formed between population biobanks and research applicants. More importantly, the existence of a consent form does not dictate the type of flow of the reciprocal exchange. It is rather the opposite, for the *seriate*, nonconditional nature of the flow will help researchers determine how best to approach the consent process as a whole while keeping in mind the protection of participants.

Finally, the examples presented above may all be categorized under "manifest" reciprocal exchanges, as they are tangibly performed or envisaged as part of the population biobank—participant relationship. A second category, which I will discuss briefly below, relates to a more concealed form of reciprocal exchange between population biobanks and participants. I will refer to these exchanges as "abstruse." Because of their relevantly different nature, I conceive of them in a separate category.

6.3.3.2 Abstruse returns

Abstruse returns are neither as clear nor as tangible as those outlined above. They nevertheless deserve some consideration, principally because they can be seen to have a growing prominence in the literature. In fact, a number of studies have revealed instances in which participants, simply in the act of participating in the research project, subjectively believe that they are receiving something in return (see Fig. 6.6). The returns to which I refer are strictly personal in nature and do not tangibly affect participant health [3].

In fact, a recent study of biobank participants has shown that some felt that donating data and samples was *itself* of personal value to them: "the thought of just sitting, waiting for the disease to take over seemed very alien. And so I thought the only proactive thing that I could do about the disease was maybe to take part in any research" [3]. According to study investigators, the quoted participant did not

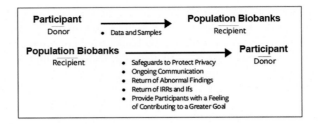

FIGURE 6.6

Participant—population biobanks reciprocal relationship.

derive any direct health benefit, yet "an alternative form of personal benefit is evident, in helping her make sense of a distressing situation" [3]. Another participant linked her participation to karma: "And this kind of karma may come back and protect me. I know that's all spooky nonsense […] At various points in my past, I've needed help, and at some point in the future I may need help" [3]. For these participants, enrolling in a biobank project allowed them to feel they have contributed to realizing a greater goal. This feeling, however, is subjective. In a 2014 paper, Stjernschantz Forsberg et al. also suggest that research participants enroll in scientific research for their own self-interest [61]. Stjernschantz Forsberg et al. argue that it is in the interest of individuals to participate in biobank research in order to allow for the greater representation of genes in future personalized medicine treatments [61]. I believe, however, that the understanding of certain participants of their enrollment in research as a way to satisfy their own sense of personal realization requires additional evidence if we are to fully appreciate its scope, rationale, and impact on participants and limitations. More importantly, the level at which the population biobank needs to act is unclear. If we are to consider this to be a return, would the population biobank need to succeed only so that participants can feel they are part of some great realization, or must it produce impactful results so that participants themselves can value from it in the future? The answer is unclear. For this reason, I have categorized this kind of return under the "abstruse" category.

As for the conception of reciprocity with which it can be identified, I argue that it is reciprocity for mutual respect. If we were to say, for the sake of argument, that the return is one in which the biobank will need to succeed and/or produce impactful results, it would be both fitting and sufficient. It would be fitting because such return would be a good and would be considered both as a good and as a return by the participant. It is also sufficient because it acknowledges the donation of participants by—in the terms used by research participants themselves—invigorating their sense of purpose. Further, by succeeding in their mandate and producing results from the use of collected data and samples, population biobanks are fulfilling what they have promised participants during recruitment. Doing so is a great manifestation of respect toward the other and a productive way to ensure that trust is maintained between all parties. The resulting reciprocal exchange is individual in nature, for it concerns only the research participant. It is also generalized in scope, as no commensurate economic return is predicted. Given that it is not negotiated, but rather personal, I consider its flow to be seriate and its value to be expressive—emphasizing positive returns for the participant and a feeling of being valued and respected.

One of my main criticisms of the individualistic conception of autonomy is its lack of acknowledgment of multilateral relationships necessarily implicated in population biobanking research, including those that involve the public and broader research community. As demonstrated in Chapter 4, relational autonomy represents a conception that can not only palliate the shortcomings of individualistic autonomy but can conceivably be adapted to the population biobank setting as well. To do so, I argued that relational autonomy must be operationalized through reciprocity. The concept of reciprocity highlights the various relationships stakeholders involved in population biobanks have between each other and reflects the vehicle in which we may better adapt the relational conception of autonomy to this research endeavor.

This section presented three types of relationships, all implicating the population biobank. Table 6.1 below summarizes their nature and characteristics. The most important kind of relationship, of course, and the one under scrutiny when discussing the disclosure of information during the consent process, is the population biobank—participant relationship. Through the concept of reciprocity, this section managed to showcase intertwining connections between this relationship and those that include both the public and research community. In fact, in the population

Table 6.1 Summary of the reciprocal relationships in population biobanks.

Relationship	Characteristics of the reciprocal exchange
Population biobank—public *Population biobank as the recipient and the public as the donor* *Population biobanks as the donor and the public as the recipient*	• conception: reciprocity for mutual respect (fittingness and sufficiency) • nature: communal • scope: generalized • flow: seriate • value: symbolic (expressive)
Population biobank—research community *Population biobank as the donor and the research community as the recipient*	• conception: reciprocity for mutual benefit (fittingness and proportionality) • nature: individual • scope: nonspecialized • flow: negotiated • value: instrumental
Population biobank—participant *Population biobank as the recipient and the participant as the donor*	• conception: reciprocity for mutual respect (fittingness and sufficiency) • nature: communal • scope: generalized • flow: seriate • value: symbolic (expressive)

biobank—public relationship, one of the mechanisms of return to the public included implementing an efficient access system for the use of data and samples donated by individual participants. In the population biobank—research community relationship, one of the mechanisms of return by the population biobank to the research participant included ensuring that strict security safeguards that protect the data and samples of participants were put in place. Correlatively, decisions made by research participants can affect both the public and the research community. This is certainly the case given that the mechanism for any potential return by the population biobank is powered by the data and samples of participants. For example, should the participant decide to limit the kinds of research projects accessing his or her data and samples, certain members of the research community might be left out and will not be able to apply for access. Similarly, if the research participant requires that he or she consents to specific future use of their data and samples as they come, delay could ensue, which would likely slow the translation of knowledge acquired from research into meaningful results for society and future generations (as seen in Chapter 4). Practically, this means that any decision participants make in the course of the consent process and throughout their participation in population biobanking would affect the public and research community. Such interconnectivity indicates that research participants are relational beings whose interests and decisions can be shaped and influenced by their connections to others. Only reciprocity allows us to fully appreciate this reality.

Given that my overarching objective is to propose a conception of autonomy that can be suitably adapted to population biobanks and ultimately respected by researchers, the next logical step would be to describe how a reciprocity-based relational autonomy affects the disclosure of information by researchers during the consent process. The following section (IV) aims to do exactly that. Moreover, I will also demonstrate the value added by considering reciprocity as a basis for relational autonomy in considerations surrounding the disclosure of information to participants. This also means that I will need to examine the limitations of adopting such conception and how these limitations can be palliated. I will turn my attention to these important questions in Section 6.5.

6.4 Reciprocity-based relational autonomy in population biobanks: how does it affect the disclosure of information to participants?

Early in this book, I set out to examine the duty to inform of researchers as portrayed in Canadian court decisions. In doing so, I described how two leading court decisions, *Halushka* and *Weiss*, emphasized the importance of full disclosure in nontherapeutic research (including all of the facts, opinions, and probabilities that needed to be presented to participants) by characterizing the duty to inform in these contexts as the most exacting possible [62,63]. I also noted that this duty was mainly centered on considerations that relate to participants in abstraction to other potential stakeholders in the research setting. I highlighted how this singular focus on participants is especially challenging at a moment in which research endeavors are becoming increasingly longitudinal (analyzed and accessed over time), international (crossing boundaries and legal jurisdictions) [64,65], and less directly focused on individuals.

In order to better understand the rationale supporting the standard set by the courts, and because the duty to inform of the researcher is correlated with the autonomy of the participant, I decided to examine the specific conception of participant autonomy that is at the heart of these decisions. After examination, I determined that the conception of autonomy that has shaped the courts' standard in the disclosure of information was an individualistic conception of autonomy influenced by liberal individualism. More importantly, I argued that this conception of autonomy presupposes a unidirectional relationship toward the participant that marginalizes the ways in which decisions can be shaped by their connections to other important stakeholders, namely the public and the research community in the population biobanking setting.

I set out to find a more appropriate conception of autonomy; one that could form the basis of a reconceived duty to inform for researchers and a new standard of disclosure in the case of complex, ongoing, and multilateral relationships established by population biobank projects. To that end, I identified relational autonomy as a potential fertile conception on which to build. I argued that in order to operationalize relational autonomy in population biobanking, the relations at the heart of this conception can only be understood through the prism of reciprocity.

This brings us to an examination of what the disclosure of information grounded on the respect of a reciprocity-based relational autonomy would look like in the context of population biobanking. More specifically, it will be important to understand how this reconceived standard of disclosure would vary in its configuration from the exacting standard set by the Courts in Canada. This will be done through a study of the level and intensity of information participants can expect to receive from population biobank researchers according to a reciprocity-based relational conception of autonomy.

First and foremost, reciprocity-based relational autonomy will require that relationality be acknowledged in any discussions between the participant and the population biobank researcher as part of their duty to inform. The participant will need to understand how they are engaged in a reciprocal relationship with the population biobank and how the population biobank is also engaged in reciprocal relationships with the public and the broader research community. Doing so will ensure that participants understand their role and any potential duties toward others. In contrast to the individualistic conception of autonomy, the decision-making process that is crystallized during the disclosure of information will not solely focus on participants but will acknowledge how decisions made by the participant may affect other stakeholders. In doing so, however, the participant should not suffer any harm or loss of autonomous choice. They will simply need to understand the interests of others—in our case, the public and the research community—and take them into account before making an autonomous decision.

Secondly, in order to ensure a meaningful discussion where participants receive proper information from the population biobank researcher, several novel considerations will play an important role in setting a reconceived duty to inform for researchers during the consent process. Chief among them should be that reciprocity-based relational autonomy must encourage the population biobank researcher to see the participant as embedded in a web of reciprocal relationships. This will allow the researcher to critically reflect with the participant on the needs and interests of other stakeholders (for example, the public and research community) and how decisions made by the participant during the consent process and throughout participation might affect them. More practically speaking, much of the intensity of the discussion should focus on the reciprocal nature of the relations these various stakeholders have within the population biobank. This is crucial because, as I showed in the previous section, nearly all of the relevant relationships revolve around the data and samples provided by the research participant. Rather than endeavor (despite the many limitations of doing so) to provide the participant with a full disclosure of all the facts, opinions, and probabilities, the disclosure of information to participants by researchers should focus on what the provision of such data and samples actually means for all of the stakeholders. When it comes to the public, participants should be informed that the use of their data and samples will allow the translation of scientific findings into scientific knowledge that will benefit society and future generations. More specifically, participants should be informed that the use of their data and samples will allow for the publication of academic works that will contribute to science and general knowledge. In order to facilitate this, the participant must know the roles played by the research community and its members. Indeed, participants should be informed that their data and samples will be accessed by future researchers who will, in turn, enrich data held at the biobank and accelerate the translation of research knowledge into clinical outcomes. In contrast to broad consent approaches studied in Chapter 3 of this book, participants are not only informed that their data and samples will be used for future unspecified research, but they will also be told why this is the case and will be informed of the intended outcome. This allows participants to clearly situate their role within the multiple existing relations and better understand their potential duties toward others. In other words, by being informed that decisions they might make will affect the public and research community, they will consider the interests of these stakeholders in their reflection. This is necessary, because doing otherwise might imperil population biobanks. As I stressed in earlier chapters, any decision that participants make in the course of the consent process and throughout their participation could very well affect the public or the research community. Such interconnectivity indicates that research participants are relational beings whose interests and decisions can be shaped and influenced by their connections to others. Given the important role played by the public and the research community in population biobanking, not considering their interests could hinder the accomplishment of the population biobank and likely slow the translation of knowledge acquired from research into meaningful results for society and future generations.

This, of course, does not mean that researchers should not consider the interests of participants. Far from it. Thanks to the respect of a reciprocity-based relational autonomy by researchers, participants will clearly understand the reciprocal relationship they have with the population biobank. While they donate their data and samples, they should expect different types of "returns" from the biobank. Such returns should be presented as aiming to protect the participant's interests, but also as a way for the population biobank to recognize the value of their donation. To that end, the population biobank researcher should inform the participant of the ways their data and samples will be protected once they are collected, when they are stored, and when they are in use by other researchers. They should be familiarized with the safeguards in place to do so. They should be informed that they will receive abnormal findings should any be identified. If they are interested, and if it is possible to do so, participants should be informed that they can also receive analytically valid, clinically significant, and actionable IRRs and incidental findings. Finally, any limitation in undertaking any of these returns should be identified and participants should be informed of ways this can be palliated. For example, in contrast with the traditional duty to inform that requires full disclosure of all facts, opinions, and probabilities, participants can be informed that some information may not be available at the time of recruitment—such as who will access their data and samples and for what purposes. One way of palliating this is to emphasize the importance of ongoing communication for the purposes of informing participants via public registries the identities of researchers who are accessing their data and samples and for what reasons. Again, the goal must be to allow participants to situate themselves within the web of relations encircling the population biobank endeavor.

In sum, a duty to inform of researchers that is rooted in the respect of a reciprocity-based relational autonomy has a number of observable characteristics: (1) it sees participants as subject to multiple relationships; (2) it ensures that they are aware of that reality and that they are informed of how their decisions may affect other stakeholders; and, (3) it explains the reciprocal nature of the relationship created between the participant and the population biobank, including ways in which the interests of participants will be protected and how their donation will be valued and recognized by the population biobank. In contrast to an approach that requires researchers to provide full disclosure of information they might not have (specific consent) or an approach that simply informs participants that their data and samples will be used by future researchers provided mechanisms of protection and ongoing communication (broad consent), the disclosure of information that I propose contains elements of broad consent, but provides a stronger conceptual grounding based on reciprocal relationships and interactions that ultimately aim to value the participant's contribution. This is a different dynamic than that found in current approaches, one that I argue is more suitable in the context of population biobanking. The disclosure of information that respects a reciprocity-based relational autonomy exhibits advantages as well as limitations. Highlighting these requires further examination, which will be the subject of the next section.

6.5 Respecting a reciprocity-based relational autonomy in the disclosure of information to participants: advantages and limitations

One prominent limitation facing population biobank researchers when they inform participants about the research project in which they are about to enroll in is the inability to foresee all potential future use of stored data and samples that are collected and stored as a result of participation. Indeed, as various uses are likely to emerge in the future, adhering to the traditional duty to inform rooted in individualistic autonomy would require a specific model of consent, one in which future uses are expected to be defined from the onset of the project. If they are not, participants would need to be reconsented every time a new use is requested by a researcher applying for access to the stored data and samples. A duty to inform that requires such full disclosure of all facts, probabilities, and opinions is difficult to satisfy in such a situation, especially given that it is largely unfeasible [66] and even potentially distressing [67] for multiple iterations of reconsent to be administered to participants over time.

Against this backdrop, neither Canada's federal government nor any of the provinces have enacted legislation specifically regulating biobanks. In contrast, a number of other jurisdictions have recognized the limitations of the model of specific consent and have proposed a pragmatic form of information provision by researchers. Alternative pragmatic models, such as broad consent, have been proposed in several international documents and were also heavily featured in the literature. However, "solution-oriented" approaches are often overly practical in nature and seem to palliate symptoms (multiple reconsent, full disclosure), rather than the cause of the problem. The problem in this case, I argue, is the individualistic conception of autonomy at the heart of the disclosure of information standard as constructed by the Canadian Courts. I argue that the individualistic conception of autonomy does not sufficiently value the roles society and the research community play in population biobanking. As a result, it centers its concern in what the participant needs to know solely for their own interests. Individualistic autonomy is thereby incapable of conceiving of a role for research participants as part of a multilateral web or relations that include other stakeholders. In that respect, the resulting traditional duty to inform of researchers is limitative.

Faced with this challenge, I have applied the concept of reciprocity to the relational conception of autonomy in order to find a more suitable conceptual basis for the disclosure of information by population biobank researchers during the consent process. I have aimed to understand the reciprocal relationships between population biobanks and participants, population biobanks and the public, and population biobanks and the research community. With these relationships in mind, I demonstrated that all four categories of biobank stakeholders feature in an intertwined web of relations, in which each thread (i.e., stakeholder) affects the others. This helped to shape a new way of looking at the duty to inform of researchers that I believe has the ability to acknowledge and sustain the multilateral relationships implicated in population biobanking research, without compromising the rights of research participants. I also believe it contains a number of other advantages that deserve a closer look (Section 6.5.1) and a few limitations that will need to be addressed should we be interested in introducing it in practice (Section 6.5.2).

6.5.1 Respecting a reciprocity-based relational autonomy in the disclosure of information to participants: advantages

I argue in this section that respecting reciprocity-based relational autonomy in the disclosure of information to participants has two main advantages other than expediency in accounting for all of the stakeholders implicated in population biobanking as shown in previous sections. First, the resulting disclosure of information helps to manage the sometimes contradictory altruistic and self-interested considerations that have been found to exist in numerous studies of the motivations of biobank participants and can help researchers better navigate them. Secondly, the standard of disclosure that is created will be more conducive to research studies and will allow for meaningful disclosure of information. I will discuss both of these advantages in turn.

6.5.1.1 Reconciling altruism and self-interest considerations

The first advantage of the proposed standard of disclosure of information is its capacity to better equip researchers in reconciling contradictory participant motivations for biobank enrollment, which are sometimes based on altruism and sometimes grounded in self-interest (the expectation of receiving something in return). As alluded to earlier in Chapter 6, certain authors have alluded to the role of altruistic motivations in the willingness of participants to carry costs in the assistance of others [68,69]. Another line of reasoning, contrastingly, identifies self-interest as a strong catalyst for participation [61]. This is a contradiction—some have even identified it as a paradox [69]. Reciprocity-based relational autonomy, when at the heart of the considerations surrounding the disclosure of information to participants, provides a conceptual basis for making sense of this paradox, reconciling self-interested and altruistic motivations in the way information is conveyed to participants wishing to enroll in biobank research.

Indeed, in accounting for the array of relationships that feature in the population biobank setting and by differentiating between them, researchers will be able to explain to participants the altruistic nature (i.e., for the benefit of others) of their enrollment in population biobanks. More specifically, participants will be informed that by providing data and samples, they will help the research community increase its understanding of disease for the benefit of society. They will also be informed that they operate within a framework of communal reciprocity between the population biobank and the public. Such discussion will help the participant understand that their decision to participate encompasses altruistic considerations. Similarly, researchers will also be able to emphasize to participants the importance of seeing their enrollment as part of a reciprocal relationship between them and the population biobank, where health-related or personal returns will be made to both value and acknowledge the contribution they have made. Such emphasis will, in turn, help researchers to efficiently manage any self-interested expectations participants may have when they are informed about their participation in population biobanks during the consent process. In contrast, the traditional duty to inform as construed by Canadian courts, by being focused on providing full disclosure to participants (which have focused centrally on the risks of participating), does not offer a comprehensive framework in which the contradictory considerations participants entertain can be nuanced and reconciled in the population biobanking context. By allowing the participant to clearly situate their role within the multiple existing relations and better understand their potential duties toward others, the resulting reconceived duty to inform anchored in the respect of a reciprocity-based relational autonomy provides researchers with more expedient means to manage participant expectations when informing them about their participation during the consent process.

6.5.1.2 From full *disclosure to* meaningful *disclosure*
This section aims at showing how reciprocity-based relational autonomy is conducive to *meaningful* disclosure of information to participants.

Earlier chapters in this book have demonstrated that the traditional duty to inform of researchers is anchored in an individualistic conception of autonomy. The resulting standard of disclosure is exacting, owing largely to the requirement that full disclosure be provided to participants without giving full consideration to the potential limitations faced by researchers in doing so. As we have seen in Chapter 3, full information might not be readily available in the context of population biobanks. The traditional duty to inform implies that biobanks, in these cases, ought to reconsent participants every time a new applicant requests data or samples if they were initially unknown or their research project was unspecified at the time of initial consent. As a result, this approach risks impeding population biobank operations with the concern that some set of samples may be used and others may not [70]. This situation will be especially likely to materialize if participants who are contacted for reconsent to a new use under the requirements of individual autonomy refuse to provide such consent or do not respond to the request. This reality is not without consequence for participants themselves. By participating in population biobank research, participants are contributing data and samples for future, unspecified research. Once these data and samples are stored, biobanks usually have an obligation to make them available to the broader research community. The goal is to increase statistical power as a way of generating useful results in the promise that such results will translate into the advancement of knowledge [14] for the benefit of society [15] and future generations. Impeding this translational mechanism by applying potentially burdensome procedures based on individualistic concerns risks alienating the crucial role of the research community and failing to generate public benefit. More importantly, however, a focus on individualistic concerns ultimately means that population biobanks fail, as I indicated above, to respect promises made to participants during recruitment and consent. This is why I argue that individualistic autonomy renders the very act of consenting and receiving information somewhat devoid of meaning. Indeed, what is the point of consenting to a research project if, in order to ensure that a participant is fully informed (for the sake of being fully informed), there is a risk of impeding the same project the participant is consenting to? In other words, what is the purpose of informing the participant of their enrollment in a research project that will likely not produce the expected outcome it was created to produce in the first place? If the disclosure of information to a participant aims at allowing them to make an informed decision about participation, will the participant truly be able to do that if they are unaware of their role within the complex web of relations found in population biobanks and how the success of the project as well as the interests of other stakeholders affect and are affected by their decisions as a participant? Full disclosure, as construed by Canadian courts, is more concerned with perfunctorily providing participants with information (even if it means reconsenting them and impeding the very research they are consenting to) than it is with meaningfully informing participants about the importance of their contribution in a way that can fulfill precisely what was promised to them.

Reciprocity-based relational autonomy, in turn, and the reconceived duty to inform associated with it do not have the perfunctory disclosure of information as a goal, but rather aim at providing participants with meaningful information throughout the consent process. The disclosure of information anchored in reciprocity-based relational autonomy aims to sufficiently acknowledge participant contributions and show respect for their donation [71]. Practically speaking, the disclosure of information grounded in a reciprocity-based relational conception of autonomy will not require researchers to provide *all* facts, opinions, and probabilities in a unidirectional and routine matter, but will instead focus the researcher's obligation on providing participants with all known information while taking into account the diverse relations that could shape their decisions. Additionally, participants will be informed of how they interact with others in the context of their participation in the research project. When crystallized in the consent process, this approach would include language on the general objectives of the population biobank (for example, benefitting the public and future generations) and the importance of sharing data and samples with the research community in pursuit of such objectives. Also included will be a discussion of the ways in which privacy and the confidentiality of data and samples will be protected, a brief overview of the overall access governance system (and attendant safeguards), and how participants may be kept informed of the identities of researchers given access to the biobank. While some of this information is currently available in certain consent forms of the population biobanks reviewed in Chapter 2, the crucial difference in approach lies in the context in which it is provided. In consent forms currently in use, some of this information is provided simply to inform the participant of the procedures in place. Looked at from the perspective of a reciprocity-based relational autonomy, this information will be provided as a way to highlight how the population biobank will express the value of participant contributions and how the population biobank will reciprocate accordingly. Additionally, by respecting a reciprocity-based relational conception of autonomy, researchers must provide a clear explanation of why participants cannot continually be updated and why they will not be asked to provide additional consent every time a request for access is processed. These explanations will turn on such issues as processing delays the risk of adversely affecting the integrity of the data. Current consent forms, such as those reviewed in Chapter 2, do not include clauses of this nature. Further, researchers must give a clear description of any potential return to the participant and how it will materialize (for example, abnormal findings, IRRs, and incidental findings). In conclusion, researchers would not cease to have responsibilities to keep participants informed, but the reconceived duty to inform of researchers will not be undertaken cursorily, but will be executed in a meaningful manner that meets the expectations of research participants who donate their data and samples and expect to know how they will be used, what the limitations are in doing so, to whom these data and samples will contribute and what the population biobank will do to both acknowledge and value their participation. Doing so reinforces the participant's autonomy as they will be able to better understand their role within the research project and the overall impact of their contribution on all the relevant stakeholders. More importantly, they will be informed of the responsibilities of others toward them and the possible returns they can expect. This will build their confidence and provide them with a threshold from which to measure outcomes in accordance with their goals and stated preferences.

Finally, the resulting standard of meaningful disclosure of information I am defending would promote the fundamental goals of population biobanks, such as encouraging the creation of generalizable knowledge for the benefit of future generations and regularly communicating research results to the public. The successful implementation of this sort of regime would not only positively empower research participants by setting aside space for recognizing their contributions, but it would also be of benefit to the population at large. I maintain that respecting a reciprocity-based relational conception of autonomy when disclosing information to participants during the consent process will nurture and sustain a multilateral, trust-based relationship between the population biobanks, researchers, participants and the public owing to the successful accomplishment of their joint endeavor, or at least, the lack of apparent barriers hampering such joint success. This will give real meaning to their decision to participate. By accounting for all of the relevant interests and relationships implicated in biobanking research, the proposed standard of disclosure of information would also encourage a more effective and efficient research paradigm, thereby hastening the translation of basic knowledge into the clinic.

6.5.2 Limitations of respecting a reciprocity-based relational autonomy in the disclosure of information to participants

Against this backdrop, I predict that the introduction of a new standard of disclosure of information grounded in a reciprocity-based conception of relational autonomy would face several limitations. Two are theoretical, while the remaining others are practical. The first theoretical limitation turns on the robustness of the notion of reciprocity for mutual respect as a basis for relational autonomy and whether it may be conflated with solidarity, a concept that has also been proposed in the literature. The second limitation addresses the perception that reciprocity-based relational autonomy may favor population biobank researchers over participants.

The third (practical) limitation involves the absence of a direct legislative foothold through which the disclosure of information anchored in reciprocity-based relational autonomy could be realized. The final limitation turns on the general applicability of this new standard of disclosure in other kinds of research projects. To better examine these limitations, the following subsections will be presented in the form of questions to which I respond. In each of my answers, I first acknowledge the relevant limitation and suggest how it may be palliated.

6.5.2.1 Is reciprocity-based relational autonomy a conceptual easy way out and how does it differ from the concept of solidarity?

My overarching argument in this book is that reciprocity-based relational autonomy is a more appropriate conception to respect when considering the disclosure of information to participants in the case of complex, ongoing, and multilateral relationships established by population biobank projects. It could be said, however, that a conception of reciprocity based on mutual respect—which I identify in various reciprocal relationships between the different stakeholders involved in population biobanks—is an overly simplistic solution to the challenges posed by an individualistic conception of autonomy. More precisely, it may be argued that reciprocity for mutual respect presumes that the relevant stakeholders in population biobank research all share mutual feelings of respect. In light of strong evidence indicating that the public remains somewhat weary about scientific research [72], some might argue that such a presumption could be somewhat naïve. If it is the case that the public has become "sensitised by various biomedical research controversies" [73], it cannot easily be presumed that a concept based intrinsically on respect will be capable of replacing such feelings while also successfully addressing the shortcomings of individual autonomy. Before addressing these considerations directly, I must first clarify several points.

First, it is critical to point out that the concept of reciprocity that I invoke as a basis for relational autonomy is not exclusively based on relationships of mutual *respect*. Indeed, as I have shown in this chapter, the relationship between population biobanks and the research community, for example, is based on a conception of reciprocity for mutual *benefit*. Only the relationships population biobanks have with participants and the public should be conceived as based on reciprocal relationships of mutual *respect*.

Secondly, "respect" in this context, based on a characterization given by Christine Hartley, refers to "respect for someone who contributed to one's project" [12] and is best described as a form of recognition of a given contribution. As I have shown in detail in this chapter, reciprocity for mutual respect does not simply take the form of a "thank you" but will rather be exhibited in a panoply of actions embodying a spirit of reciprocity. In the relationship between a population biobank and the public, where the public is a donor receiving a return from the biobank, such return is typically realized in the form of efficient access mechanisms, the return of general results (newsletters), and the academic dissemination of findings. Each of these activities aims to recognize the public's contributions, which, as I have said, are usually offered in the form of public funds. Returns as efficient access mechanisms, for example, aim to ensure that public money is not wasted. As I argued in Chapter 4, facilitating efficient access to data and samples by researchers will help to streamline benefits society derives from biobanking research. In turn, ongoing communication on the part of the biobank keeps the public informed of ongoing research and encourages transparency. Scientific publications, somewhat distinctly, are more tangible, short-term milestones that contribute to the foundational goals of the research enterprise: improving health outcomes for the benefit of society and future generations. High-quality publications allow the research community to benefit from advances in knowledge and facilitate the translation of research findings to the clinic.

The notion of respect lies at the core of the relationship between population biobanks and participants and better reflects the dynamic that underlies these relationships. There is no need to reiterate the kinds of return extended to participants on the model of reciprocity, but I would nevertheless like to emphasize that respect, understood as a principle, has been a central consideration in numerous statements on the ethical conduct of research since the *Nuremberg Code* [74]. For example, the *Tri-Council Policy Statement*, a recent ethics guidance document, states that "[r]espect for [p]ersons recognizes the intrinsic value of human beings and the respect and consideration that they are due" [75]. It goes on to say that respect for persons includes a moral obligation to respect their autonomy [75]. This is directly related to what I am proposing in this dissertation: a reciprocity-based relational conception of autonomy that lies at the heart of the duty to inform of researchers during the consent process.

A recent conceptual study concluded that the value of respect in research is largely communal in nature [67]. Interestingly, on the topic of the future use of biosamples, the same study suggests that limiting consent (and by extension, the disclosure of information) to the use of data and samples for a single project, in which participants will be asked to reconsent for each subsequent new use, will likely "increase the costs associated with research, reduce the data or samples available, compromise data integrity, or may even cause unwarranted harm to participants or their families" [67]. Beyond that, the study argues that respect would not be best promoted under an approach that demands subsequent reconsent. I have argued that continuous reconsent would be required according to an individualistic conception of autonomy. Reciprocity for mutual respect, however, offers a powerful argument against this practice.

Another point of contention revolves around whether the reciprocity for mutual respect conception can be conflated with the notion of solidarity present in recent literature in the biobanking field. Such conflation may be understandable given the many times reciprocity and solidarity have been presented in unison [76]. To examine this point, I briefly turn to literature describing the concept of solidarity in biobanking and demonstrate how reciprocity and solidarity are, ultimately, distinct. Further, I will take the opportunity to briefly comment on the concept of solidarity as an approach to palliate the limitations of the individualistic conception of autonomy.

As early as 2001, the concept of solidarity was under discussion as a potential alternative approach to the ethics of informed consent in the biobanking context [77]. This concept was introduced using arguments similar to those presented by authors who have suggested there is a duty to participate in research. In this case, this proposed duty of research was framed by a concept of solidarity:

> *it could be argued that one has a duty to facilitate research progress and to provide knowledge that could be crucial to the health of others. This principle of solidarity would strongly contradict a view that no research should be conducted if it would not directly benefit those participating in a study [77].*

More recently, the concept of solidarity has been considered in the context of biobank governance more generally. Like reciprocity, the concept of solidarity has been proposed in a range of different settings, from labor [78] to migration [79] and economics [80] to biomedicine [81]. Popularized in the field of biomedical ethics by Barbara Prainsack, Alena Buyx, and others, solidarity draws on critical scholarship to challenge individual autonomy's flawed foundational premise that individuals are fundamentally independent and atomistic [81]. Instead of "treating social reality as reducible to the actions of independent individuals," solidarity focuses on relations and interactions [81]. Prainsack and others view solidarity not primarily as a value or sentiment, but as a *practice*, one in which individuals express a willingness to accept economic, emotional, or social costs "to assist others with whom they recognize similarity in a relevant respect" [81]. Prainsack and Buyx have defended the role of solidarity in understanding the willingness of persons to accept costs and assist others [68,76]. Their departing premise, which mirrors my own, is that contemporary reliance on individual autonomy encounters a number of problems [68,76]. Biobank governance regimes, according to these authors, are fixated on the protection of autonomy and the avoidance of risk. In their 2011 report *"Solidarity: Reflections on an Emerging Concept in Bioethics"* [76], Prainsack and Buyx distinguish between solidarity and reciprocity. In their view, reciprocity refers to symmetrical arrangements of giving and receiving in which "what one gives and one receives is equal in value (not in kind)" [76]. According to this definition, reciprocity differs from solidarity to the extent that acts of solidarity are not dependent on the receipt of something in return, or even on the expectation of receiving something in return [76].

While I maintain that reciprocity for mutual respect and solidarity are indeed distinct concepts, I believe they are linked insofar as reciprocity can be understood—in certain cases—as a *cause* of solidaristic action. In fact, as I have examined earlier in this book, generalized exchanges—which are often associated with the conception of reciprocity as mutual respect—are oriented toward maintaining social solidarity and are on the high-end spectrum of solidarity-building varieties of reciprocity [76]. When the public donates, for example, they are not entrenched in a relationship that will require commensurable return, but are more interested in improving the health and well-being of fellow members of society. This is an ideal example of a reciprocal exchange within a conception of reciprocity for mutual respect that leads to solidaristic action.

Before I turn to addressing the second potential theoretical limitation of introducing reciprocity-based relational autonomy in population biobanks, I should also briefly comment on how Prainsack and Buyx contemplate the relationship between autonomy and solidarity, given that I have introduced reciprocity as a basis for relational autonomy earlier in this chapter in order to see whether there are elements that can be useful to my overall analysis.

Prainsack and Buyx describe three distinct varieties of solidarity [68]. The first variety, which they identify as Tier 1, operates at the level of the individual. People act in solidarity, the authors argue, when they believe that they are similar to others in some relevant respect [68]. Tier 2, in contrast, is a kind of group-level solidarity. Here, acts of solidarity between individuals are so widespread that they become shared community practices [68]. The authors suggest that this type of solidarity is more institutionalized than Tier 1, but is not yet consolidated by legal arrangements. The authors identify Tier 3 as institutionalized solidarity, in which acts of solidarity are entrenched in "contractual relationships or hard law" [68]. Health insurance and public pension systems are prominent examples [68]. Prainsack and Buyx defend the introduction of a multitiered conception of solidarity by appealing to the proposed need to move away from the dominant focus on autonomy in biobank governance. They criticize the restrictive interpretation of autonomy in contemporary medical ethics as synonymous with consent [68] and criticize unnecessarily exacting efforts to protect participants from relatively small risks:

> *focusing our efforts and resources to protect participants from these small risks leads to barriers for research. Significant resources are currently used for (re-) consenting procedures and formal risk prevention requirements (e.g., obtaining new research ethics approval for a slightly modified research question, recontacting, and reconsenting participants) [68].*

From this view, it is clear that Prainsack and Buyx share my perspective that individualistic autonomy is an unstable foundation for population biobanking. While solidarity is conceived as a response to individualistic autonomy, reciprocity differs insofar as it operates as a conceptual compliment to an alternative to individualism: *relational autonomy*. While solidarity emphasizes the acceptance of costs incurred to help others with whom an individual feels a degree of connection, reciprocity primarily emphasizes the acceptance of costs in exchange for a *benefit bestowed*. Although solidarity has clearly a great deal to contribute to our current thinking in biomedical ethics, reciprocity offers a solid foundation for population biobanking insofar as actors in population biobanking are motivated by a web of conveyed benefits and interconnected relations. Reciprocity also captures the dynamics of conveyed benefits that I think are essential for complementing relational autonomy and more effectively grounding the duty to inform in population biobanking.

I now turn to the second theoretical limitation, this one focusing on whether reciprocity-based relational autonomy favors population biobank researchers over participants.

6.5.2.2 Does a reciprocity-based relational autonomy favor population biobank researchers over participants?

Another potential limitation of reciprocity-based relational autonomy is the perception that it would favor researchers over participants. This is perhaps to be expected given that relational autonomy has itself been subject to this same criticism [82,83]. The more direct question here is this: will the introduction of reciprocity-based relational autonomy infringe individual rights for the sake of giving preference to the rights of other actors? The answer, in my view, is that it will not. In contrast to the individualistic conception of autonomy, the decision-making process that is generated by the new disclosure of information standard I propose does not solely focus on participants, but rather simply acknowledges the manner in which decisions made by the participant might affect other stakeholders.

Further, under the framework I postulate, the existing rights of participants will be upheld, the privacy and confidentiality of their data and samples will be protected. and they will continue to be informed about the use of their data and samples over time. Nothing in reciprocity-based relational autonomy aims to withhold information that is known at the time of participant consent. In fact, if we take the web of relations that forms the basis of relational autonomy, seen through the lens of reciprocity, we can understand how the participant's interests are necessarily upheld. In public—population biobank reciprocal relationships, one of the returns identified above (with the public as donor) is the creation of an efficient access mechanism to the data and samples of participants. As we have seen in Chapter 2 of this book, the creation of an efficient access mechanism requires that the mechanisms in place accord with ethical principles. Efficient access involves not only the development of required documentation but also the formation of bodies tasked with evaluating and approving access requests [13]. In essence, biobank participants have agreed to have their data and samples used in future, yet-unspecified research projects. This kind of agreement necessitates the creation of mechanisms that ensure the process is carried out in a way that respects the wishes of participants (as expressed in consent forms) and protects both their privacy and the confidentiality of their data and samples [84]. In the reciprocal relationship between the population biobank and the research community, one of the returns undertaken by the researchers that access data and samples is the implementation of strict security safeguards throughout the use of the data and samples as a way of ensuring that the reidentification of participants or unauthorized data and sample access is avoided [40,85].

Lastly, in the reciprocal relationship between the population biobank and participants, the interests of participants are considered at multiple levels. While I will not reiterate them in detail here, they are worth mentioning briefly. The first level at which the interests of participants are considered is through the implementation of procedures to protect the privacy of participants during the collection, long-term storage, and the sharing of data and samples with the research community. Furthermore, if an abnormal finding (such as high blood pressure) were to be identified during the assessment and recruitment stage, the participant would be informed and, if needed, provided access to emergency medical services. In the longer term, where an IRR or IF that is analytically valid, clinically significant, and actionable is found, the participant would be informed. To be sure, they would not be informed in cases where consent had not been given or where such disclosure is impracticable.

Reciprocity-based relational autonomy does not favor population biobank researchers over participants, but rather aims at continuing to respect and protect them while offering them a more meaningful disclosure of information. The relevant disclosure of information allows for the fulfillment of what has been promised to participants: a better understanding of the causes of chronic disease and the factors that influence health and illness across the Canadian population for the benefit of society and future generations [86]. In doing so, the participant does not suffer harm and their autonomy will be strengthened as their decision will be based on more comprehensive and meaningful information. They will also be asked to understand that the interests of others, such as the public and research community, are implicated in their decision-making and should be taken into account when their autonomy is exercised.

6.5.2.3 Would a new standard for disclosure of information to participants grounded in reciprocity-based relational autonomy have a legislative foothold?

Jurists responding to new social challenges presented by technological innovation may, out of habit, be drawn to legislation as a first mode of recourse. Importantly, however, autonomy in research has been a subject of debate in biobanking for nearly 2 decades [87]. A number of jurisdictions have decided to legislate (for example, Belgium [88] and Estonia [89]), while others, such as Canada, have not. The decision whether to provide legislated response has been characterized by some as the "Collingridge dilemma" [90]. On the one hand, if a state responds with legislation, there is a risk that such legislation will quickly become outdated in the face of rapid technological advancement. On the other hand, if legislation is not enacted, there is a risk that the technology will become so entrenched that it will no longer be easily amenable to regulatory oversight [90].

That said, a situation in which the proposed new standard of disclosure/duty to inform, anchored in reciprocity-based relational autonomy, is not incorporated into legislation might raise the worry that the model is thereby unenforceable. But this, I think, is overstated. I am generally skeptical that enacting legislation (*hard law*) is necessarily the best approach to biobank oversight. This is so for several reasons. In general, biobank regulation should aim to both protect participants and present specific guidance to biobanks. In terms of participant protection, already-enacted legislation on privacy, confidentiality, and research integrity, among others, should provide acceptable levels of legal oversight. Existing legal regimes, moreover, may be complemented by specific guidance for biobanks, in the form of *statements* and *policies*, sometimes referred to as *soft law*. I will briefly discuss these documents and demonstrate how they enjoy legal weight even if they are not legislative in nature.

Policies and *statements* represent what I call a "bottom-up approach" to biobank regulation. They are developed and adopted by grass-roots organizations that include a panoply of stakeholders, such as researchers or participants that have decided to tackle specific issues and deliberate within their communities with the purpose of proposing guidance relevant to their own fields. The Réseau de médecine génétique appliquée du Québec is a good example of such an organization. Over the years, it has published a number of guidelines for the genetic research community [91]. Such documents become legally meaningful to the extent that they are adopted and followed by the research community. In fact, where the law is "unclear or incomplete, the court will often refer to nonlegal professional instruments to make legal findings. In these cases, the judge will usually invoke a policy, code, or guideline with expert testimony to determine whether it represents customary practice" [92]. Although this excerpt is presented in the negligence context, I believe it is nevertheless pertinent here and will apply for soft law generally. In the absence of hard law, policy documents tend to have a "significant, perhaps even decisive, impact on a judge's conclusion. This essentially results in the professional community, rather than the legislator or the court, determining the legal standard of care" [92]. This is especially true in the case of widely adopted documents such as the *Tri-Council Policy Statement*. Compared with legislation, these documents have the distinct advantage of being substantially more flexible. This is so in the sense that they do not typically require lengthy and politically motivated amendment processes, as hard law inevitably does. The implication is that such notions as the proposed new standard of disclosure/duty to inform anchored in reciprocity-based relational autonomy may be incorporated into standards of practice with substantially less difficulty than its enactment by the legislator. An organization that represents researchers in population biobanking or participant advocates may introduce the proposed new standard in policy documents, explain its merits, rationale, and characteristics and, in turn, adopt it for use by its members and the broader research community. With time, documents of this kind are applied and relied upon in the research setting and thereby take on considerable legal weight in the absence of hard law.

6.5.2.4 Can the new standard of disclosure of information Be applicable to projects other than population biobanks?

A fourth potential limitation relates to the possible difficulty of applying my proposed model in contexts other than population biobanks. I will first look at the use of the new standard of disclosure in other kinds of biobanks before examining whether it can fruitfully be used in other kinds of research. When looking to other kinds of biobanks as examples, it becomes clear that a standard of disclosure of information anchored in reciprocity-based relational autonomy may also apply. Where there is a biobank, after all, there will invariably also be a research community. This community will interact with the relevant biobank in much the same way as they would interact with a population biobank. As far as the public is concerned, the specific characteristics of these relationships will depend on the specific biobank's objectives. For example, if the biobank has a disease-specific purview, then the relevant public stakeholders will largely consist of persons who suffer from the disease in question. But considering that the aim of research, generally conceived, is to provide generalizable knowledge that will be translated into better health outcomes for future generations, the public remains necessarily implicated in the web of research relations.

The participant—biobank relationship, moreover, will be reciprocal whether the research project is disease specific or not. All that may end up subject to modification is the *purpose* of the relationship of reciprocity between the actors. It may, for example, change from reciprocity for mutual respect to reciprocity for mutual benefit. Using a conception of reciprocity for mutual benefit, for example, might be feasible in situations where the biobank has a therapeutic aim [3], which may create an expectation among participants that they will derive some healthcare benefit from enrollment. With these examples in mind, there will always be a need to account for the reciprocal character of relationships between biobanks and stakeholders.

Outside of the biobanking context, however, the applicability of a new standard of disclosure might not be as obvious. For one thing, the relevant stakeholders would be quite different. Secondly, the implicated research community would be distinct in relevant ways in contexts where sample and data storage and provision are not part of the research process. That said, the public will still likely have a vital role to play. This is so because, as I have said above, the ultimate goal of health research is always the same: to provide generalizable knowledge and to improve future health outcomes.

In my view, it is far from certain that reciprocity will serve as an applicable modus operandi outside of the biobanking context. The requirement, however, that the contributions of research participants be acknowledged and respected— which is at the heart of the concept of reciprocity—should, nonetheless, be a value worth advancing in a variety of research settings. The appeal of that value, I submit, is a sufficiently universal aspiration in research relationships that will transcend biobanking.

6.6 Conclusion

Over the past 2 decades, much has been written about the legal and ethical issues associated with biobanking, and about those associated with population biobanks in particular. Crucial considerations, such as privacy, data, and sample access and the return of research results and IFs, have been extensively debated in the academic literature. A wide range of potential solutions and responses to such challenges have been proposed. As I described in Chapter 3, Section 3.4, issues of consent have been a similarly foundational concern, which has taken on special and pronounced prominence in the field of population biobanks. For example, the literature has grappled with questions about whether specific consent is acceptable and whether broad consent respects requirements of law. Proposed solutions, however, have often been problematic, largely because they tend to lack theoretical rationale and serve mainly practical purposes that ultimately aim at accommodating biobanks in the face of ongoing changes in research culture. Yet, despite obvious limitations facing population biobanks when it comes to providing information to participants in the current Canadian legal system [93], nothing has been written about the main thrust of the problem: that our current conception of autonomy is individualistic in nature and, as a consequence, fails to acknowledge the multilateral relationships necessarily implicated in population biobank research. With these problems in mind, I built on a relational conception of autonomy by complementing it with the concept of reciprocity.

A reciprocity-based relational autonomy is preferable to liberal individualism to the extent that the former includes all of the relevant stakeholders in its analysis and appropriately describes the nature of their interactions. In fact, by agreeing to take part in population biobank research, participants contribute their data and samples to future, unspecified scientific study. Once data and samples are stored, the imperative function of the biobank is to make them available for use by the broader research community. This occurs with the goal, as I have said, of increasing statistical power in order to generate more scientifically useful results. In turn, such results generate meaningful knowledge [14] for the benefit of society [15] and future generations. Ultimately, this works to improve population health and increase public trust in science [94]. Ultimately, this chain of function shows us that interactions based in the data and samples of a participant necessarily, as a matter of design, implicate a range of actors and interests. Put another way, participants are but one element in a larger environment that requires multiple stakeholders working in tandem. Only reciprocity can reflect this reality and operationalize the relational conception of autonomy.

Reciprocity is not, as seen in Chapter 5, itself a novel concept. It has been presented in a number of economics, sociological, and medical analyses. That being said, its application in relational autonomy forms the basis of a reconceived duty to inform and a new standard of disclosure of information that respects and protects research participants while providing them with a meaningful opportunity to exercise the said autonomy. The resulting disclosure of information sees the participant as embedded within multiple relations; it ensures that participants are aware of that reality and that they are informed of how their decisions can affect other stakeholders (namely the public and research community). In contrast to an approach that requires researchers to provide full disclosure of information they might not otherwise have (based on an individualistic conception of autonomy) or an approach that simply informs participants that their data and samples will be used by future researchers, the proposed standard of disclosure provides a better context for the sharing of information based on reciprocal relationships and interactions that ultimately aim to value the participant's contribution and benefit future generations. This is why I argue that the proposed standard of disclosure of information grounded in reciprocity-based relational autonomy is more appropriate in the population biobanking context.

References

[1] Knoppers BM, Chadwick R. Human genetic research: emerging trends in ethics. Nat Rev Genet 2005;6(1):75–6.

[2] Gottweis H, Gaskell G, Starkbaum J. Connecting the public with biobank research: reciprocity matters. Nat Rev Genet 2011;12(11):738–9.

[3] Locock L, Boylan AMR. Biosamples as gifts? How participants in biobanking projects talk about donation. Health Expect 2016;19(4):805–16.

[4] Hobbs A, Starkbaum J, Gottweis U, Wichmann H, Gottweis H. The privacy-reciprocity connection in biobanking: comparing German with UK strategies. Public Health Genomics 2012;15(5):272–84.

[5] Critchley C, Nicol D, McWhirter R. Identifying public expectations of genetic biobanks. Publ Understand Sci 2017;26(6):671–87.

[6] Pellegrini I, Chabannon C, Mancini J, Viret F, Vey N, Julian-Reynier C. Contributing to research via biobanks: what it means to cancer patients. Health Expect 2014;17(4):523–33.

[7] Nicol D, Critchley C. Benefit sharing and biobanking in Australia. Publ Understand Sci 2012;21(5):534–55.

[8] Kanellopoulou N. Reconsidering altruism, introducing reciprocity and empowerment in the governance of biobanks. In: Kaye J, Stranger M, editors. Principles and practice in biobank governance. Farnham: Ashgate; 2009. p. 33–52.

[9] Kanellopoulou N. Reciprocity, trust, and public interest in research biobanking: in search of a balance. In: Lenk C, et al., editors. Human tissue research: a European perspective on the ethical and legal challenges. New York: Oxford University Press; 2011. p. 45–53.

[10] Eriksen KÅ, Sundfør B, Karlsson B, Råholm M-B, Arman M. Recognition as a valued human being: perspectives of mental health service users. Nurs Ethics 2012;19(3):357. cited in Sima Sandhu et al. Reciprocity in therapeutic relationships: A conceptual review (2015) 24 International J Mental Health Nursing. p. 464.

[11] CanPath Portal. Leadership and governance. 2020. https://canpath.notadev.site/governance/ [Accessed 21.03.23].

[12] Hartley C. Two conceptions of justice as reciprocity. Soc Theor Pract 2014;40(3):409–32.

[13] Shabani M, Knoppers BM, Borry P. From the principles of genomic data sharing to the practices of data access committees. EMBO Mol Med 2015;7(5):507–9.

[14] OECD. Guidelines on human biobanks and genetic research databases, best practice 4.1. 2009. www.oecd.org/science/biotechnologypolicies/44054609.pdf [Accessed 21.03.11].

[15] Human Genome Organisation (HUGO). Principles agreed at the first international strategy meeting on human Genome sequencing. 1996. https://web.ornl.gov/sci/techresources/Human_Genome/research/bermuda.shtml#1 [Accessed 21.03.24].

[16] Borry P, Bentzen HB, Budin-Ljøsne I, Cornel MC, Howard HC, Feeney O, et al. The challenges of the expanded availability of genomic information: an agenda-setting paper. J Community Genet 2018;9(2):103–16.

[17] Becker LC. Reciprocity. Chicago: The University of Chicago Press; 1993.

[18] Macneil IR. Exchange revisited: individual utility and social solidarity. Ethics 1986; 96(3):581–93.

[19] Molm LD, Schaefer DR, Collett JL. The value of reciprocity. Soc Psychol Q 2007; 70(2):199—217.

[20] CARTaGENE. Second wave information brochure for participants. 2014. https://cartagene.qc.ca/sites/default/files/documents/consent/cag_2e_vague_brochure_en_v3_7apr2014.pdf [Accessed 21.03.11].

[21] Atlantic PATH. Consent and brochure (obtained through correspondence). 2013. p. 2—4.

[22] The Tomorrow project. Alberta: Study Booklet; 2011. p. 4—6 (obtained through correspondence).

[23] CARTaGENE Twitter Account. https://twitter.com/_cartagene_?lang=en. (Accessed 21.03.23).

[24] CanPath Portal. Access policy. 2020. https://canpath.ca/wp-content/uploads/2020/08/Access_Policy_Approved_July_29_2020_final.pdf [Accessed 21.03.24].

[25] New York Times. For women, confusion about alcohol and health. Parker-pope T. 9 Oct. 2007. https://well.blogs.nytimes.com/2007/10/09/at-cocktail-time-shots-of-confusion/ [Accessed 21.03.23].

[26] EurekAlert!. Kaiser permanente study: alcohol amount, not type—wine, beer, liquor—triggers breast cancer. 27 Sept. 2007. https://www.eurekalert.org/pub_releases/2007-09/kpdo-kps092207.php [Accessed 21.03.23].

[27] Li Y, Baer D, Friedman GD, Udaltsova N, Shim V, Klatsky AL. Wine, liquor, beer and risk of breast cancer in a large population. Eur J Cancer 2009;45(5):843—50.

[28] Awadalla P, Boileau C, Payette Y, Idaghdour Y, Goulet JP, Knoppers B, et al. Cohort profile of the CARTaGENE study: Quebec's population-based biobank for public health and personalized genomics. Int J Epidemiol 2013;42(5):1285—99.

[29] Molm LD. The structure of reciprocity. Soc Psychol Q 2010;73(2):119—31.

[30] Burton PR, Hansell AL, Fortier I, Manolio TA, Khoury MJ, Little J, et al. Size matters: just how big is BIG? Quantifying realistic sample size requirements for human genome epidemiology. Int J Epidemiol 2009;38(1):263—73.

[31] CanPath Portal. Access policy, s 12. 2020. https://canpath.ca/wp-content/uploads/2020/08/Access_Policy_Approved_July_29_2020_final.pdf [Accessed 21.03.24].

[32] CanPath Portal. Access policy, s 9. 2020. https://canpath.ca/wp-content/uploads/2020/08/Access_Policy_Approved_July_29_2020_final.pdf [Accessed 21.03.24].

[33] CanPath Portal. Access policy, s 8 a. 2020. https://canpath.ca/wp-content/uploads/2020/08/Access_Policy_Approved_July_29_2020_final.pdf [Accessed 21.03.24].

[34] Becker LC. Reciprocity, justice, and disability. Ethics 2005;116(1):9—39.

[35] CanPath Portal. Access policy, s 15. 2020. https://canpath.ca/wp-content/uploads/2020/08/Access_Policy_Approved_July_29_2020_final.pdf [Accessed 21.03.24].

[36] CanPath Portal. Data and samples access application form. 2016. https://portal.partnershipfortomorrow.ca/agate/register/#/join [Accessed 21.03.23].

[37] CanPath Portal. Publications policy. https://portal.canpath.ca/sites/live-7x35-1-release-1597681236-mr-portal/files/CanPath%20Publications%20Policy%20PDF%20Version.pdf. (Accessed 21.03.24).

[38] Milanovic F, Pontille D, Cambon-Thomsen A. Biobanking and data sharing: a plurality of exchange regimes. Genom Soc Pol 2007;3(1):17—24.

[39] CanPath Portal. Access policy, ss 7b & 13. 2020. https://canpath.ca/wp-content/uploads/2020/08/Access_Policy_Approved_July_29_2020_final.pdf [Accessed 21.03.24].

[40] CanPath Portal. Access policy, s 6. 2020. https://canpath.ca/wp-content/uploads/2020/08/Access_Policy_Approved_July_29_2020_final.pdf [Accessed 21.03.24].

[41] CanPath Portal. Access agreement. 2016. https://portal.partnershipfortomorrow.ca/agate/register/#/join [Accessed 21.03.24].

[42] CanPath Portal. Access agreement, clause 5.4.1. 2016. https://portal.partnershipfortomorrow.ca/agate/register/#/join [Accessed 21.03.24].

[43] CanPath Portal. Canadian partnership for tomorrow project. 2018. www.partnershipfortomorrow.ca [Accessed 21.03.11].

[44] CanPath Portal. Approved projects. 2015. https://portal.partnershipfortomorrow.ca/mica/research/projects [Accessed 21.03.23].

[45] Zawati MH, Rioux A. Biobanks and the return of research results: out with the old and in with the new? JL Med Ethics 2011;39(4):614–20.

[46] Zawati MH, Knoppers BM. International normative perspectives on the return of individual research results and incidental findings in genomic biobanks. Genet Med 2012;14(4):484–9.

[47] Beskow LM, Burke W, Fullerton SM, Sharp RR. Offering aggregate results to participants in genomic research: opportunities and challenges. Genet Med 2012;14(4):490–6.

[48] Ravitsky V, Wilfond BS. Disclosing individual genetic results to research participants. Am J Bioethics 2006;6(6):8–17.

[49] Bredenoord AL, Kroes HY, Cuppen E, Parker M, van Delden JJ. Disclosure of individual genetic data to research participants: the debate reconsidered. Trends Genet 2011;27(2):41–7.

[50] Solberg B, Steinsbekk KS. Managing incidental findings in population based biobank research. Norsk Epidemiologi 2012;21(2):195–6.

[51] Knoppers BM, Dam A. Return of results: towards a lexicon? JL Med Ethics 2011;39(4):577–82.

[52] Wolf SM, Crock BN, Van Ness B, Lawrenz F, Kahn JP, Beskow LM, et al. Managing incidental findings and research results in genomic research involving biobanks and archived data sets. Genet Med 2012;14(4):361–84.

[53] Canadian Institutes of Health Research, Natural Sciences and Engineering Research Council of Canada & Social Sciences and Humanities Research Council of Canada. How to address material incidental findings: guidance in applying (TCPS 2) (2018). Ottawa: Secretariat on Responsible Conduct of Research; 2019. art 3.4.

[54] Canadian Institutes of Health Research, Natural Sciences and Engineering Research Council of Canada & Social Sciences and Humanities Research Council of Canada. How to address material incidental findings: guidance in applying TCPS2 (2018). Ottawa: Secretariat on Responsible Conduct of Research; 2019 [Glossary: "Impracticable"].

[55] Knoppers BM, Deschênes M, Zawati MH, Tassé AM. Population studies: return of research results and incidental findings policy statement. Eur J Hum Genet 2013;21(3):245–7.

[56] Ontario Health Study. Consent form (obtained through correspondence). 2014.

[57] Canadian Alliance for Healthy Hearts and Minds. Participant information and consent sheet (Thunder Bay site). (obtained through correspondence).

[58] Canadian Alliance for Healthy Hearts and Minds. Policy on managing the return of severe structural abnormalities. (obtained through correspondence).

[59] Anand SS, Tu JV, Awadalla P, Black S, Boileau C, Busseuil D, et al. Rationale, design, and methods for Canadian alliance for healthy hearts and minds cohort study (CAHHM) - a Pan Canadian cohort study. BMC Public Health 2016;16:650.

[60] Canadian Institutes of Health Research, Natural Sciences and Engineering Research Council of Canada & Social Sciences and Humanities Research Council of Canada. Tri-council policy statement: ethical conduct for research involving humans [TCPS 2]. Ottawa: Secretariat Responsible for the Conduct of Research; 2014. art. 3.3 ("Consent shall be maintained throughout the research project. Researchers have an ongoing duty to provide participants with all information relevant to their ongoing consent to participate in the research").

[61] Forsberg JS, Hansson MG, Eriksson S. Why participating in (certain) scientific research is a moral duty. J Med Ethics 2014;40(5):325−8.

[62] Solomon WC. RJQ 731 at 743, 48 CCLT 280 (QCSC) para 89. 1989.

[63] Philips-Nootens S, Lesage-Jarjoura P, Kouri RP. Éléments de responsabilité civile médicale. In: Le droit dans le quotidien de la médecine. 4 ed. Cowansville: Yvon Blais; 2017.

[64] Knoppers BM, Zawati MH. Population biobanks and access. In: Canestrari S, Zatti P, editors. Il governo del corpo: Trattato di biodiritto, vol. 2. Milan: Giuffrè Editore; 2011. p. 1181.

[65] Taylor K. Paternalism, participation and partnership—the evolution of patient centeredness in the consultation. Patient Educ Counsel 2009;74(2):150−5.

[66] Tassé AM, Budin-Ljøsne I, Knoppers BM, Harris JR. Retrospective access to data: the ENGAGE consent experience. Eur J Hum Genet 2010;18(7):741−5.

[67] Pieper IJ, Thomson CJ. The value of respect in human research ethics: a conceptual analysis and a practical guide. Monash Bioeth Rev 2014;32(3):232−53.

[68] Prainsack B, Buyx A. A solidarity-based approach to the governance of research biobanks. Med Law Rev 2013;21(1):71−85.

[69] Nobile H, Vermeulen E, Thys K, Bergmann MM, Borry P. Why do participants enroll in population biobank studies? A systematic literature review. Expert Rev Mol Diagn 2013;13(1):35−47.

[70] Jones KH, Laurie G, Stevens L, Dobbs C, Ford DV, Lea N. The other side of the coin: harm due to the non-use of health-related data. Int J Med Inf 2017;97:43−51.

[71] Canadian Institutes of Health Research, Natural Sciences and Engineering Research Council of Canada & Social Sciences and Humanities Research Council of Canada. Tri-council policy statement: ethical conduct for research involving humans TCPS 2. Ottawa: Secretariat Responsible for the Conduct of Research; 2014. Chap. 2.

[72] Johnsson L. Trust in biobank research. Uppsala: Uppsala University; 2013. p. 51. http://www.irdirc.org/wp-content/uploads/2018/02/Trust-in-biobank-research.pdf [Accessed 21.03.24].

[73] Caulfield T. Biobanks and blanket consent: the proper place of the public good and public perception rationales. King's LJ. 2007;18(2):209−26.

[74] Nuremberg Military Tribunals. Permissible medical experiments. In: Trials of war criminals before the Nuremberg military tribunals under control council law, vol 10:2. Washington, DC: US Government Printing Office; 1949.

[75] Canadian Institutes of Health Research, Natural Sciences and Engineering Research Council of Canada & Social Sciences and Humanities Research Council of Canada. Tri-council policy statement: ethical conduct for research involving humans (TCPS 2). Ottawa: Secretariat Responsible for the Conduct of Research; 2014. Chap. 1.

[76] Prainsack B, Buyx A. Solidarity: reflections on an emerging concept in bioethics, xiii. London: Nuffield Council on Bioethics; 2011. Available from: http://nuffieldbioethics.org/wp-content/uploads/2014/07/Solidarity_report_FINAL.pdf.

[77] Chadwick R, Berg K. Solidarity and equity: new ethical frameworks for genetic databases. Nat Rev Genet 2001;2(4):318—21.

[78] Prainsack B, Buyx A. The value of work: addressing the future of work through the lens of solidarity. Bioethics 2018;32(9):585—92.

[79] Tazzioli M, Walters W. Migration, solidarity and the limits of Europe. Global Discourse 2019;9(1):175—90.

[80] Morandeira-Arca J, Etxezarreta-Etxarri E, Azurza-Zubizarreta O, Izagirre-Olaizola J. Social innovation for a new energy model, from theory to action: contributions from the social and solidarity economy in the Basque Country. Innovat Eur J Soc Sci Res 2021:1—27.

[81] Prainsack B. The "We" in the "Me": solidarity and health care in the era of personalized medicine. Sci Technol Hum Val 2018;43(1):21—44.

[82] Christman J. Relational autonomy, liberal individualism, and the social constitution of selves. Phil Stud 2004;117(1/2):143—64.

[83] McLean SA. Autonomy, consent and the law. London: Routledge; 2009.

[84] Lemmens T, Austin LM. The end of individual control over health information: promoting fair information practices and the governance of biobank research. In: Kaye J, Stranger M, editors. Principles and practice in biobank governance. Farham: Ashgate; 2009. p. 243—51.

[85] CanPath Portal. Access portal documents, which include a data access policy, a publications policy, an intellectual property policy and a data access application form. https://portal.partnershipfortomorrow.ca/request-access. (Accessed 21 03 23).

[86] BC Generations Project. Consent form. British Columbia (obtained through correspondence). 2014. p. 3—5.

[87] Laurie G. Reflexive governance in biobanking: on the value of policy led approaches and the need to recognise the limits of law. Hum Genet 2011;130(3):347—50.

[88] Loi relative a l'obtention et a l'utilisation de matériel corporel humain destiné a des applications médicales humaines ou a des fins de recherche scientifique (Belgium) M.B. 30/12/2008. https://www.ieb-eib.org/fr/pdf/l-20081219-rech-mater-humain.pdf. (Accessed 21 03 11).

[89] Human Genes Research Act 2000 (Estonia) RT I (104, 685). Available from: https://www.riigiteataja.ee/en/eli/531102013003/consolide. (Accessed 21 03 12).

[90] Collingdridge D. The social control of technology. London: St. Martin's Press; 1982. p. 58.

[91] Réseau de médecine génétique appliquée du Québec (RMGA). Network Appl Genet Med 2016. https://www.rmga.qc.ca/?set_lang=eg/issues [Accessed 21.03.24].

[92] Campbell A, Glass KC. The legal status of clinical and ethics policies, codes, and guidelines in medical practice and research. McGill LJ 2000;46:473—85.

[93] Caulfield T, Murdoch B. Genes, cells, and biobanks: yes, there's still a consent problem. PLoS Biol 2017;15(7):2—6.

[94] Shabani M, Borry P. You want the right amount of oversight: interviews with data access committee members and experts on genomic data access. Genet Med 2016;18(9):892—7.

General conclusion

In a paper on biobanking consent, authors noted that the "biobanking community needs to come to terms with [...] the reality that the types of consent used in biobanking often do not meet the requirements necessitated by relevant legal norms" [1]. Use of the term, "types of consent" refers in this case to an array of practical consent solutions, such as broad or tiered consent, that biobanks have adopted, and which deviate from the traditional specific consent model. In principle, I agree with the spirit of their proposition. It is certainly the case that biobanking challenges the traditional consent model founded in the relevant legal norms. Having said that, I am concerned that this debate on the provision of information to participants has been conducted rather superficially. The view defended by the authors focuses primarily on symptoms, namely, the limitative characteristics of specific consent and the patchwork of deficient solutions biobanks have proposed. From there, the authors assess the legal and ethical implications of these various available approaches. This analysis, in my view, would benefit from starting with a different perspective. It would be far more fruitful, I think, to begin the analysis at the heart of the duty to inform of researchers. More precisely, since specific consent is the crystallization of a certain way of approaching the duty to inform of researchers, we might first consider how it is theoretically justified. Are there, for example, any limitations embedded in requirements set by the courts? Rather than focus on how practical solutions fare when evaluated against these requirements, we should also consider what the requirements *ought to be* in the first place.

In this book, I set out to answer this set of questions using population biobanking as a case model. Doing so, I endeavored to meet several objectives. First, I aimed to explicate the present jurisprudential interpretation of the duty to inform of researchers in Canada. Underpinning this assessment, I developed an understanding of the correlative conception of autonomy courts have applied as a way of justifying the relevant standard associated with the dominant model of the duty to inform. To this end, I traced in Chapter 1 the evolution of the duty to inform during the 20th century in Canada. I showcased how researchers must presently conduct themselves in a way that respects an individualistic conception of autonomy when informing their participants about research participation. This state of affairs, I argued, has an outsized negative impact on population biobanking and on the relevant duties of researchers. More concretely, I showed how courts determined that the duty is substantially more exacting in the research context than it is in the clinic. Respect for autonomy, as it has been conceived by the courts, demands that researchers fully disclose all facts, opinions, and probabilities to participants when recruiting for participation in research. Often, such disclosure is impractical or otherwise simply impossible.

Building on this analysis, I turned to the second objective, namely, to examine limitations of the individualistic conception of autonomy in the context of population biobanking. This required several stages of inquiry. First, in Chapter 2, I laid out numerous unique characteristics of population biobanks and differentiated them from alternative ways of conducting health research. I engaged in a qualitative document analysis of internal documents shared with Canadian biobank participants. This analysis revealed that the public and the research communities play a central and critical role in this species of research. Second, aligned with this review of the characteristics of population biobanks, I developed a tangible understanding of the practical and theoretical limitations of the individualistic conception of autonomy in the population biobanking context. In Chapter 3, I focused specifically on the practical limitations by drawing on the consent forms and associated documents reviewed in Chapter 2. Parallel with this effort, I also reviewed policies, guidelines, and statements that have addressed the duty to inform of researchers in population biobanks. I described how population biobanks are constitutionally unable to foresee every possible use of stored data and samples. This impossibility means that they must systematically deviate from the requirement of full disclosure of all facts, probabilities, and opinions required in Canadian law. Further, Chapter 3 similarly demonstrated the infeasibility, as the individualistic conception of autonomy would require, of reconsenting participants every time a new project requests access to a biobank's data and samples.

While Chapter 3 discussed shortcomings of the individualistic conception of autonomy from a practical perspective, I took a more theoretical approach in Chapter 4. There, I demonstrated that the individualistic conception of autonomy is unable to account for the complexities of benefit considerations in the research setting. From there, I established that the individualistic conception of autonomy, with its unidirectional focus on the participant, is an implausible grounding for the disclosure of information by researchers during the consent process in population biobanks. This is so primarily owing to the multilateral relationships that are necessarily and fundamentally implicated in population biobank research, and, in particular, in projects involving the broader research community and the public at large.

Against this backdrop, my third and final objective was to propose an alternative conception of autonomy that would respond to the practical and theoretical limitations of the individualistic conception of autonomy identified in Chapters 3 and 4. After considering solutions that have been proposed in the literature, I determined that most are unsuited to the population biobanking context. One clear exception, however, was uncovered: the relational conception of autonomy. With relational autonomy's promise being noted, I argued that in order for this conception to be adapted to the population biobanking context, it must first be complemented by a concept that reflects the specific relations and interests engaged by these projects. The concept capable of doing this work, I proposed, was reciprocity.

In an attempt to better understand the concept of reciprocity, I examined numerous proposed theories of reciprocity in Chapter 5. More precisely, I explored potential reciprocal exchanges by outlining their nature, scope, flow, and value. I similarly demonstrated that there exists two dominant conceptions of reciprocity in the literature: reciprocity for mutual benefit and reciprocity for mutual respect. By establishing the contours of reciprocity, it became possible to apply this concept to relational autonomy and to propose novel parameters for the disclosure of information by researchers in population biobanking. This was the function of Chapter 6. By identifying the kinds of relations existing between various stakeholders using the prism of reciprocity, I demonstrated that reciprocity offers an appropriate and plausible grounding for relational autonomy in population biobanks. It does so, despite certain limitations, because of its capacity to both acknowledge and sustain the multilateral relationships implicated in population biobanking research. This is accomplished, notably, without compromising the rights of research participants. Owing to this understanding of how reciprocity grounds relational autonomy, the consequent reconceived duty to inform of researchers considers research participants as embedded within a web of relations. Reciprocity ensures that participants are meaningfully informed of existing relationships in the research project and are aware of how the decisions they make may affect other stakeholders, including the public and research community. Contrasted with individualistic conception of autonomy that demands that researchers provide full disclosure of information, or the practical accommodation in which participants are simply informed that samples will be used in future unspecified research, the proposed reciprocity-based standard of disclosure provides a more convincing framework for sharing information with participants during the consent process. An approach based on reciprocal relationships and interactions ultimately aims, I argued, at demonstrating the value of a participant's contribution and to benefit future generations.

In light of this analysis, a number of interconnected findings loom with particular prominence. First, the modern research landscape is complex and varied. Clinical trials, like those in *Halushka* and *Weiss*, are only one kind of research study. A one-size-fits-all approach, in which the oversight taken in one species of research is reflexively transplanted in others, should generally be avoided (unless the underlying principles are universal). Crucially, this does not imply that requirements for autonomy and the disclosure of information must be individually tailored to specific research projects. Instead, it means that the principles underlying such requirements should be founded, as far as possible, on denominators common across research settings. Reciprocity, in my view, does exactly that. By concretizing relationships grounded in the acknowledgement and respect for contributions made by research participants, the concept of reciprocity advances values capable of transcending clinical trials and biobanking.

Second, while individual participants are certainly an important part of the research infrastructure, there are not its singular focus. As I have demonstrated in this book, a theory that focuses unilaterally on research participants would tend to alienate other critical stakeholders. Of course, this in no way indicates that we should adopt a model that would infringe or ignore the rights of participants. Rather, I am suggesting only that, at a moment in which health research is becoming increasingly observational and less focused on individuals, it is becoming critical to strike a balance between the protection of the interests of participants and those of other important stakeholders in the research enterprise without compromising the interests of participants.

Third, the standard of disclosure of information by researchers should not be assessed by its intensity, but rather primarily by how meaningful it is. The issue of determining whether informed consent is truly informed has been a ceaseless refrain in the literature [2—4]. One of the reasons for this turns on the ways researchers have carried out their duties with the aim of providing full disclosure to research participants. The exacting requirements set by the Canadian courts are typically communicated in consent forms that are dozens of pages in length. But the researcher's provision of information should not be primarily guided by *how much* information is provided, but rather, *how meaningful* the information is to the participant. This means that research participants should be provided with information that helps them understand their overall participation and role within the research endeavour as well as how they contribute to it. Conceiving of the disclosure of information simply in terms of intensity is one of the major tribulations we have inherited from both *Halushka* and *Weiss*. In those cases, the duty to inform of researchers was found to be more exacting, that is, higher in intensity than that which exists in the clinical setting. Treating the clinical and research duties to inform as varying in nature only as far as their intensity is concerned is inapt. They should rather be treated as two different creatures. While similar in certain nontrivial ways, the relevant actors, setting, and purpose of intervention differ markedly. Relational autonomy based in reciprocity shifts our focus away from the strictures implanted in Canadian jurisprudence on the assumption that health research is largely uniform in nature, and toward recognizing the kinds of relationships and contributions engaged in population biobanking and other research endeavours.

At this point, it becomes important to consider the future of the framework I have proposed. Population biobanks are a relatively novel form of medical research, raising largely unprecedented legal and ethical issues, many of which are likely to continue to arise. It is critical that policymakers and biobank researchers stay ahead of this curve, anticipating the issues I have raised and beginning to develop research designs capable of appreciating and palliating them. Building on research undertaken in this book, I believe it is especially important to pursue the creation of reciprocal and adaptive processes in population biobanks and, in doing so, to engage all relevant stakeholders in their assessment. Access governance models that facilitate the flow of data within the research setting or as between research and the clinic could be inspired by reciprocity. We may draw on reciprocity's recognition of patient's and participant's contributions, as well as on its capacity to account for stakeholders whose interests are vitally implicated in research projects. Beyond that, template consent forms that include language reflective of the reciprocal and relational nature of these various research relationships may be drafted and distributed to population biobank researchers, research participants, and research ethics boards in order to obtain their feedback and impressions.

More substantially, we may undertake a qualitative study in which the understanding of reciprocity-based relational consent processes among research participants is gauged. As part of this process, it would be important to consider whether research participants feel their contribution is being valued by a process founded on reciprocity-based relational autonomy. Participant views in this context should be measured against perspectives in other approaches to informed consent in population biobanking (such as specific or broad consent). The difference between these and the framework I have proposed is, primarily, that the reciprocity-based relational model is based on a thorough theoretical examination and not solely on practical, reflexive solutions founded in the need to palliate superficial symptoms. The results of such research will, I believe, better inform future practices in the field of precision medicine, where longitudinal research projects promise to be the norm and where data and samples provided by research participants continue to be invaluable [5].

References

[1] Caulfield T, Murdoch B. Genes, cells, and biobanks: Yes, there's still a consent problem. PLoS Biol 2017;15(7):2—6.

[2] Ogloff JR, Otto RK. Are research participants truly informed? Readability of informed consent forms used in research. Ethics Behav 1991;1(4):239—40.

[3] Ghooi RB. Ensuring that informed consent is really an informed consent: Role of videography. Perspect Clin Res 2014;5(1):3—4.

[4] Zawati MH. Liability and the legal duty to inform in research. In: Joly Y, Knoppers BM, editors. Routledge Handbook of Medical Law and Ethics. London: Routledge; 2015. p. 199—210.

[5] Kraft SA, Cho MK, Gillespie K, Halley M, Varsava N, Ormond KE, et al. Beyond consent: building trusting relationships with diverse populations in precision medicine research. Am J Bioeth 2018;18(4):3—20.

Bibliography

Legislation

Canada

Alberta
Health Professions Act, RSA 2000, c H-7.

British Columbia
Health Care (Consent) and Care Facility (Admission) Act, RSBC 1996, c 181.
Health Professions Act, RSBC 1996, c 183.

Manitoba
The Health Care Directives Act, CCSM 1993, c H-27.

Ontario
Health Care Consent Act, 1996, SO 1996, c 2, s A.
Regulated Health Professions Act, 1991, SO 1991, c 18.

Quebec
Charter of Human Rights and Freedoms, RSQ, c C-12.
Civil Code of Québec.
Code of Ethics of Physicians, RRQ, c M-9, r 17.
Code of Ethics of Physicians, RRQ 1981, c M-9, r 4.
Professional Code, RSQ, c C-26.
Medical Act, RSQ, c M-9.

Saskatchewan
Health Care Directives and Substitute Health Care Decision Makers Act, SS 1997, c H-0.001.

Foreign Legislation

Act on Biobanks 2000 (Iceland) no. 110. as amended by Act No. 27/2008 and Act No. 48/2009.
Biobanks in Medical Care Act 2002 (Sweden).
Finnish Biobank Act 688/2012.
Human Biobanks Management Act (Taiwan), Hua-Zong-Yi-Yi-Tzu No. 09900022481.
Human Genes Research Act 2000 (Estonia) RT I (104, 685).
Law 14/2007, of 3 July, on Biomedical Research (Spain).

Other Normative Documents

American Medical Association (AMA). Principles of medical ethics. 1903. www.ama-assn.org/about/publications-newsletters/ama-principles-medical-ethics.
Australian Office of Population Health Genomics. Guidelines for human biobanks, genetic research databases and associated data. 2010. www.genomics.health.wa.gov.au/publications/docs/guidelines_for_human_biobanks.pdf.
Austria Bioethics Commission. Biobanks for medical research—amendments to the bioethics commission report of May 2007. 2011. http://archiv.bka.gv.at/DocView.axd?CobId=43808.

Bioethics Advisory Committee (BAC). Ethics guidelines for human biomedical research. Part III—Consent, sections on clinically significant incident findings and guidelines on consent. 2015. online: http://www.bioethics-singapore.org/images/uploadfile/fullReport.pdf.

Canadian Institutes of Health Research, Natural Sciences and Engineering Research Council of Canada & Social Sciences and Humanities Research Council of Canada. Tri-council policy statement: ethical conduct for research involving humans TCPS 2. Ottawa: Secretariat Responsible for the Conduct of Research; 2018.

Canadian Institutes of Health Research, Natural Sciences and Engineering Research Council of Canada & Social Sciences and Humanities Research Council of Canada. How to address material incidental findings: guidance in applying (TCPS 2) (2018). Ottawa: Secretariat on Responsible Conduct of Research; 2019.

Collège des médecins du Québec (CMQ). Le médecin et la recherche clinique. 2007. http://www.cmq.org.

Council for International Organization of Medical Sciences (CIOMS). International ethical guidelines for health-related research involving humans. 2016. https://cioms.ch/publications/product/international-ethical-guidelines-for-health-related-research-involving-humans.

Council of Europe. Convention for the protection of human rights and dignity of the human being with regard to the application of biology and medicine: convention on human rights and biomedicine. 4 April 1997. ETS No 164 (entered into force 1 December 1999).

Council of Europe. Additional protocol to the convention on human rights and biomedicine, concerning biomedical research. 25 January 2005. ETS No195 (entered into force 1 October 2007).

Deuttscherr Etthiikrratt. Human biobanks for research. Berlin: Deuttscherr Etthiikrratt; 2010.

EC. Directive 2001/20/Ec of The European Parliament and of The Council of 4 April 2001 [2001] OJ, L 212/34.

EC. Biobanks for Europe: a challenge for governance. Report of the Expert Group on Dealing with Ethical and Regulatory Challenges of International Biobank Research Luxembourg. European Commission; 2012.

Global Alliance for Genomics and Health. Framework for the responsible sharing of genomic and health-related data. 2014. www.ga4gh.org/genomic-data-toolkit/regulatory-ethics-toolkit/framework-for-responsible-sharing-of-genomic-and-health-related-data/.

Health Canada Panel on Research Ethics. Guidance for health Canada. Biobanking of Human Biological Material; 2011.

Human Genome Organisation (HUGO). Sharing data from large-scale biological research projects: a system of tripartite responsibility. 2003. www.genome.gov/Pages/Research/WellcomeReport0303.pdf.

Human Genome Organisation (HUGO). Ethics committee statement on human genomic databases. HUGO; 2002. http://www.eubios.info/HUGOHGD.htm.

Human Genome Organisation (HUGO). Ethics committee statement on DNA sampling: control and access. 1998. http://hrlibrary.umn.edu/instree/dnastatement.html.

Human Genome Organisation (HUGO). Principles agreed at the first international strategy meeting on human genome sequencing. 1996. https://web.ornl.gov/sci/techresources/Human_Genome/research/bermuda.shtml.

Human Genome Organization (HUGO). Ethics committee statement on benefit-sharing. 2000. www.who.int/genomics/elsi/regulatory_data/region/international/043/en/.

International Declaration on Human Genetic Data. UNESCOR, 32nd Sess, Resolutions, Item 22, SHS/BIO/04/1 REV. 2003.

McGill University — Faculty of Medicine. General guidelines for biobanks and associated databases. 2015. www.mcgill.ca/medresearch/files/medresearch/guidelines_for_biobanks_and_associated_databases.march2015.pdf.

National Cancer Research Institute. Biobank quality standard: collecting, storing, and providing human biological material and data for research. 2014. www.ncbi.nlm.nih.gov/books/NBK50729/.

National Institutes of Health. NIH genomic data sharing policy. 2014. online: https://grants.nih.gov/grants/guide/notice-files/not-od-14-124.html.

Network of Applied Genetic Medicine (RMGA). Statement of principles: human genomic research. 2000. Réseau de Médecine Génétique Appliquée du Québec (RMGA), www.rmga.qc.ca/admin/cms/images/large/enoncedeprincipesrechercheengenomiquehumaine_en_000.pdf.

OECD. Guidelines on human biobanks and genetic research databases. 2009. www.oecd.org/science/biotechnologypolicies/44054609.pdf.

Réseau de Médecine Génétique Appliquée (RMGA). Énoncé de principes sur la conduite éthique de la recherche en génétique humaine concernant des populations. 2003. http://research.chusj.org/getmedia/0e7575bb-7efc-4c5a-8349-a282579a16a2/BER_MDIE_enoncedeprincipes-conduiteethiquerecherche_FR.pdf.aspx.

The National Commission for the Protection of Human Subjects of Biomedical and Behavioral Research. The belmont report: ethical principles and guidelines for the protection of human subjects of research. Washington: US Government Printing Office; 1978.

Universal Declaration on the Human Genome and Human Rights. UNESCOR, 29th Sess, Resolutions, Item 16, 29 C/Res. 31. 2005.

United Nations Secretariat of the Convention on Biodiversity. Open access as benefit sharing? of benefits arising from their utilization. 2011. www.cbd.int/abs/doc/protocol/nagoya-protocol-en.pdf.

United States Department of Human Health and Services. Final revisions to the common rule. 2017. Federal Register 82:12, www.gpo.gov/fdsys/pkg/FR-2017-01-19/pdf/2017-01058.pdf.

World Medical Association. Declaration of Helsinki - Ethical principles for medical research involving human subjects. 64th WMA general assembly. 2013. Fortaleza, www.wma.net/policies-post/wma-declaration-of-helsinki-ethical-principles-for-medical-research-involving-human-subjects/.

World Medical Association. Declaration of Taipei on ethical considerations regarding health databases and biobanks. 2016. www.wma.net/policies-post/wma-declaration-of-taipei-on-ethical-considerations-regarding-health-databases-and-biobanks/.

Jurisprudence

Supreme Court of Canada

AC v Manitoba (Child and Family Services), 2009 SCC 30, [2009] 2 SCR 181.

Arndt v Smith (1997) 2 SCR 539, 148 DLR (4th) 48.

Ciarlariello v. Schacter (1993) 2 SCR 119 at 135, 100 DLR (4th) 609.

Hopp v. Lepp (1980) 2 SCR 192, 112 DLR (3d) 67.

Laferrière v Lawson (1991) 1 SCR 541, 78 DLR (4th) 609.

McInerney v MacDonald (1992) 2 SCR 138, 93 DLR (4th) 415.

Reibl v Hughes, (1980) 2 SCR 880, 114 DLR (3d) 1.
Rodriguez v British Columbia, (1993) 3 SCR 519, 107 DLR (4th) 342.
Starson v Swayze, 2003 SCC 32, [2003] 1 SCR 722.

Alberta

Cory v Bass, 2012 ABCA 136, 522 AR 220.
Dickson v Pinder, 2010 ABQB 269, 489 AR 54.
Halkyard v Mathew, 2001 ABCA 67, 277 AR 373.
Mangalji v Graham, (1997), 194 AR 116, 47 Alta LR (3d) 19 (ABQB).
Martin v Capital Health Authority, 2007 ABQB 260, 74 Alta LR (4th) 206.
Paniccia Estate v Toal, 2011 ABQB 326, 521 AR 34.
Paquette v Giuffre, 2011 ABQB 425, 512 AR 389.
Rhine v Millan, 2000 ABQB 212, 263 AR 201.
Seney v Crooks, 1998 ABCA 316, 223 AR 145.
Zimmer v Ringrose, (1981), 28 AR 69, 124 DLR (3d) 215 (ABCA).

Manitoba

Gerelus v Lim et al, 2008 MBCA 89, 231 Man R (2d) 23.
Jaglowska v Kreml, 2003 MBCA 113, 177 Man R (2d) 280.
Lyne v McClarty, 2001 MBQB 88, 155 Man R (2d) 191.
Thiessen v Hota, 2005 MBQB 248, 198 Man R (2d) 158.

New Brunswick

Doiron v Haché, 2003 NBQB 26, [2003] NBR (2d) (Supp) No 7.
Kitchen v McMullen, (1989), 100 NBR (2d) 91, 2 DLR (4th) 481 (NBCA).
Kueper v McMullin, (1986), 73 NBR (2d) 288, 30 DLR (4th) 408 (NBCA).
White v Sirois, 2009 NBQB 3, 339 NBR (2d) 373.

Newfoundland and Labrador

Brushett v Cowan, (1987), 64 Nfld & PEIR 262, 40 DLR (4th) 488 (NFSC).
Gallant v Brake-Patten, 2012 NLCA 23, 321 Nfld & PEIR 77.

Nova Scotia

Considine v Camp Hill Hospital, (1982), 50 NSR (2d) 631, 133 DLR (3d) 11 (NSSC).

Ontario

Brics v Stroz, (2002) OTC 171 (ONSC).
Jaskiewicz v Humber River Regional Hospital, (2001) 4 CCLT (3d) 98 (ONSC).
Leblanc v Hunt, 2011 ONSC 1333.
Malette v Shulman, (1990), 72 OR (2d) 417, 67 DLR (4th) 321 (ONCA).
Murphy v Langlois, (1999), 90 OTC 252 (ONSC).
Philion v Smith, (2008), 61 CCLT (3d) 113, 169 ACWS (3d) 221 (ONSC).
Pittman Estate v Bain, (1994), 35 CPC (3d) 67 (ONCJ).
Ross v Welsh, (2003) 18 CCLT (3d) 107 (ONSC).
Scardoni v Hawryluck, (2004) 69 OR (3d) 700 (ONSC).
Sterritt v Shogilev, (2009) OJ No 2063 (QL) (ONSC).
Symaniw v Zajac, (1996) 12 OTC 275 (ONCJ).

Prince Edward Island
Harris v Beck Estate, 2009 PECA 8, [2009] PEIJ No 14 (PEICA).

Quebec
Baum c Mohr, 2006 QCCS 2608.
Bécotte c Durocher, 2002 CanLII 115 (QCSC).
Bernier c Décarie, 2005 QCCA 705 (CanLII).
Bouchard c Villeneuve, (1996) RJQ 1920 (Sup Ct).
Chouinard c Landry, (1987) RJQ 1954, [1987] RRA 856 (QCCA).
Comeau c Léveillé, 1998 CarswellQue 3475 (WL Can) (Tribunal des professions).
Courtemanche c Potvin, 1996 CarswellQue 2434 (WL Can) (QCSC).
CSSS Alphonse Desjardins c, B(A), 2012 QCCS 811 (WL Can).
Drolet c Parenteau, 26 CCLT (2d) 168, [1994] RJQ 689 (QCCA).
Ferland c Ghosn, 2006 QCCS 4858, [2006] RRA 1069.
Godin c Quintal, (2002) RJQ 2925, [2002] RRA 741 (CA).
Guénard c Houle, 2010 QCCS 2628, [2010] RRA 894.
Hussul c Mitmaker, 2006 QCCS 1381, [2006] RRA 471.
Labrie c Gagnon, 2002 CarswellQue 2841 (WL Can) (QCSC).
Lalonde c Tessier, 2011 QCCS 3935 (CanLII).
Lamirande c Dumais, 2008 QCCQ 3459.
Lussier c Centre d'hébergement Champlain, 224 NR 238 (WL Can) (QCSC).
B(M) c Centre hospitalier Pierre-le Gardeur, 238 DLR (4th) 312, [2004] RJQ 792 (QCCA).
Marcoux c Bouchard, (1999) RRA 447 (QCCA).
Ménard c Archambault, 2010 QCCS 264, [2010] RRA 118.
Michaud c Gomez, (2001) RJQ 2788 (QCCA).
Pelletier c Coulombe, (1996) RJQ 2314, [1996] RRA 1237 (QCSC).
Pelletier c Roberge, 41 QAC 161, [1991] RRA 726 (QCCA).
Institut Philippe-Pinel c AG et le Curateur Public, 66 QAC 81, [1994] RJQ 2523 (QCCA).
R c Pelletier, (2004) RJQ 2608 (CQ).
Rafferty v Kuczycky, (1989) RRA 582 (QCSC).
Soltani v Desnoyers, 2008 QCCS 1720, [2008] RRA 753.
Watters v White, 2012 QCCA 257, 92 CCLT (3d) 1.
Weiss c Solomon, (1989) RJQ 731, 48 CCLT 280 (QCSC).

Saskatchewan
Baert v Graham, 2011 SKCA 21, [2011] SJ No 151.
Halushka v University of Saskatchewan, (1965), 53 DLR (2d) 436 (SKCA).

Secondary Material: Monographs
Baudouin JL, Deslauriers P, Moore B. La responsabilité civile. 8th ed., vol 2. Cowansville, Que: Éditions Yvon Blais; 2014.
Beauchamp TL, Faden RR. A history and theory of informed consent. New York: Oxford University Press; 1986.

Beauchamp TL, Childress JF. Principles of biomedical ethics. 6 ed. Oxford: Oxford University Press; 2001.

Becker LC. Reciprocity. Chicago: The University of Chicago press; 1993.

Childress JF, Mount Jr E. Who should decide? Paternalism in health care. London, England: SAGE Publications Sage UK; 1983.

Collins F. The language of life: DNA and the revolution in personalized medicine. New York: Harper Collins Publishers; 2010.

Collingdridge D. The social control of technology. London: St. Martin's Press; 1982.

Dworkin G. The theory and practice of autonomy. Cambridge: Cambridge University Press; 1988.

Hobbes T. In: MacPherson C, editor. Leviathan (1656). London: Penguin Books; 2003.

Herring J. Relational autonomy and family law. London: Springer Publishing; 2014.

Kant I. In: Hill TE, Zweig A, editors. Groundwork for the metaphysics of morals. Oxford: Oxford University Press; 2009.

Kottow M. From justice to protection: a proposal for public health bioethics. New York: Springer; 2012.

Kouri RP, Philips-Nootens S. L'intégrité de la personne et le consentement aux soins. 4th ed. Cowansville, Que: Éditions Yvon Blais; 2017.

Laurie G. Genetic privacy: a challenge to medico-legal norms. Cambridge: Cambridge University Press; 2002.

McLean S. Autonomy, consent and the law. London: Routledge; 2010.

Mayrand A. L'inviolabilité de la personne humaine. Montréal: Éditions Wilson & Lafleur; 1975.

Mill JS. In: Stillinger J, editor. Three essays. Oxford: Oxford University Press; 1975.

Monroe KR. The heart of altruism: perceptions of a common humanity. New Jersey: Princeton University Press; 1996.

Nedelsky J. Law's relations: a relational theory of self, autonomy, and law. Oxford: Oxford University Press; 2011.

Niall S, Seglow J. Altruism. New York: Open University Press—McGraw Hill; 2007.

O'Neill O. Autonomy and trust in bioethics. Cambridge: Cambridge University Press; 2002.

Orwell G. Politics and the english language. London: Horizon; 1946.

Palmer GH. Altruism: its nature and varieties. New York: Charles Scribner's Sons; 1919.

Patton MQ. Qualitative research and evaluation methods. 3d ed. Thousand Oaks: Sage Publications; 2002.

Philips-Nootens S, Lesage-Jarjoura P, Kouri RP. Éléments de responsabilité civile médicale. Le droit dans le quotidien de la médecine. 4 ed. Cowansville: Éditions Yvon Blais; 2017.

Prainsack B, Buyx A. Solidarity: reflections on an emerging concept in bioethics. Wiltshire: Nuffield Council on Bioethics; 2011.

Rawls J. Political liberalism. New York: Columbia University Press; 2005.

Robertson GB, Picard EI. Legal liability of doctors and hospitals in Canada. 5th ed. Toronto: Thomson Reuters; 2017.

Royer JC, Lavallée S. La preuve civile. 4th ed. Cowansville, Que: Éditions Yvon Blais; 2008.

Walzer M. Spheres of justice: a defense of pluralism and equality. New York: Basic Books; 1983.

Secondary Material: Chapters in Monographs

Barbara P. Biobanks: a definition. In: Mascalzoni Deborah, editor. Ethics, law and governance of biobanking. Dordrecht: Springer; 2015.

Bédard K, Wallace S, Lazor S, Knoppers B. Potential conflicts in governance mechanisms used in population biobanks. In: Kaye J, Stranger M, editors. Principles and practice in biobank governance; 2009.

Boulanger M. La réduction des risques en soins de santé: perspectives macroscopique et microscopique du patient. In: Service de la formation permanente du Barreau du Québec. Cowansville. Que: Éditions Yvon Blais; 2002.

Boulet D. Les soins de santé pour le majeur inapte: ce que la Loi ne dit pas. In: Barreau du Québec, La protection des personnes vulnérables (2012), vol 344. Cowansville, Que: Éditions Yvon Blais; 2012.

Bernheim E. Repenser la vulnérabilité sociale en termes d'égalité réelle: une contribution des droits de la personne. In: Barreau du Québec, La protection des personnes vulnérables (2011), vol 330. Cowansville, Que: Éditions Yvon Blais; 2012.

Caplan AL. What no one knows cannot hurt you: the limits of informed consent in the emerging world of biobanking. In: Solbakk H, Holm S, Hofmann B, editors. The ethics of research biobanking. London: Springer; 2009.

Charpentier M. L'hébergement des personnes âgées vulnérables: Une analyse à la croisée du social et du juridique. In: Barreau du Québec, La protection des personnes vulnérables (2012), vol 344. Cowansville, Que: Éditions Yvon Blais; 2012.

Deleury E, Goubau D. Le droit à l'intégrité physique. In: Goubau D. Le droit des personnes physiques. 4th ed. Cowansville, Que: Éditions Yvon Blais; 2008.

Dworkin G. Paternalism. In: Wasserstrom RA, editor. Morality and the Law. Belmont: Wadsworth Publishing Company; 1971.

Elger B. Consent and use of samples. In: Elger B, et al., editors. Ethical issues in governing biobanks: global perspectives. Aldershot: Ashgate Publishing; 2008.

Frank B. Réflexions éthiques sur la sauvegarde de l'autonomie. In: Barreau du Québec, Pouvoirs publics et protection (2003), vol 182. Cowansville, Que: Éditions Yvon Blais; 2003.

Ganguli-Mitra A. Benefit-sharing and remuneration. In: Elger B, et al., editors. Ethical issues in governing biobanks: global perspectives. Aldershot: Ashgate Publishing; 2008.

Giroux MT. Contrat thérapeutique et bienveillance exceptionnelle. In: Barreau du Québec, La protection des personnes vulnérables (2010), vol 315. Cowansville, Que: Éditions Yvon Blais; 2010.

Godard B. Involving communities: a matter of trust & communication. In: Einsiedel E, Timmermans F, editors. Crossing over: genomics in the public arena. Calgary: University of Calgary Press; 2005.

Hans-Martin S, May AT. Advance directives: balancing patient's self-determination with professional paternalism. In: Helmchen H, Sartorius N, editors. Ethics in Psychiatry. Netherlands: Springer; 2010.

Hansson MG. Striking a balance between personalised genetics and privacy protection from the perspective of GDPR. In: Slokenberga S, Tzortzatou O, Reichel J, editors. GDPR and biobanking: individual rights, public interest and research regulation across Europe. Cham: Springer; 2021.

Hallinan D. Biobank oversight and sanctions under the general data protection regulation. GDPR and biobanking. Cham: Springer; 2021.

Hinkley AE. Two rival understandings of autonomy, paternalism, and bioethical principlism. In: Engelhardt HT, editor. Bioethics critically reconsidered. Dordrecht: Springer Netherlands; 2012.

Hofmann B, Solbakk H, Holm S. Consent to biobank research: one size fits all? In: Solbakk H, Holm S, Hofmann B, editors. The ethics of research biobanking. London: Springer; 2009.

Kanellopoulou N. Reconsidering altruism, introducing reciprocity and empowerment in the governance of biobanks. In: Kaye J, Stranger M, editors. Principles and practice in biobank governance. Farnham: Ashgate; 2009.

Kanellopoulou N. Reciprocity, trust, and public interest in research biobanking: in search of a balance. In: Lenk C, et al., editors. Human tissue research: a european perspective on the ethical and legal challenges. New York: Oxford University Press; 2011.

Kaye J. Biobanking networks-What are the governance challenges? In: Kaye J, Stranger M, editors. Principles and practice in biobank governance. Farnham, UK: Ashgate; 2009.

Knoppers B, Abdul-Rahman MZ. Biobanks in the literature. In: Bernice E, editor. Ethical issues in governing biobanks: global perspectives. Farnham: Ashgate Publishing; 2008.

Knoppers BM, Zawati MH. Population biobanks and access. In: Canestrari S, Zatti P, editors. Il governo del corpo: Trattato di biodiritto. Giuffrè Editore, v. 2. Milan: Giuffrè Editore; 2011.

Knoppers BM, Avard D, Thorogood A. Informed consent in genetics. In: eLS. Chichester: John Wiley & Sons; 2012.

Kouri RP. Le consentement aux soins: aperçu général et quelques questions controversées. In: Chambre de notaires du Québec, Cours de perfectionnement du notariat 2011. Cowansville, Que: Éditions Yvon Blais; 2011.

La Charité R. Les droits de la personnalité. In: Barreau du Québec, Personnes, famille et successions: Collection de droit 2011–2012, vol 3. Cowansville, Que: Éditions Yvon Blais; 2011.

Lemmens T, Austin LM. The end of individual control over health information: promoting fair information practices and the governance of biobank research. In: Kaye J, Stranger M, editors. Principles and practice in biobank governance. Farham: Ashgate; 2009.

Ménard JP. Le refus catégorique de soins revu et corrigé. L'aptitude à consentir aux soins médicaux: la Cour suprême redéfinit les propositions de la Cour d'appel du Québec. In: Barreau de Québec, Famille et protection (2005), vol 219. Cowansville, Que: Éditions Yvon Blais; 2005.

Ménard JP. L'impact de la Loi sur la protection des personnes dont l'état mental présente un danger pour elles-mêmes ou pour autrui sur le consentement aux soins. In: Barreau du Québec, Développements récents en droit de la santé mentale (1998). Cowansville, Que: Éditions Yvon Blais; 1998.

Nitschmann K. Biobanks and the law: thoughts on the protection of self-determination with regards to France and Germany. In: Dabrock P, Taupitz J, Ried J, editors. Trust in Biobanking. Berlin: Springer; 2012.

Rainville A, Lafleur MC. L'absence de mécanismes de révision dans le cadre des requêtes en autorisation de traitements: une violation du principe de sauvegarde de l'autonomie? Réflexion, pistes de solutions et difficultés. In: Barreau du Québec, La protection des personnes vulnérables (2011), vol 330. Cowansville, Que: Éditions Yvon Blais; 2011.

Robertson JA. Ethical and legal issues in genetic biobanking. In: Knoppers BM, editor. Populations and genetics: legal and socio-ethical perspectives. Leiden: Brill Academic Publishers; 2003.

Shabani M, Chassang G, Marelli L. The impact of the GDPR on the governance of biobank research. In: GDPR and biobanking individual rights, public interest and research regulation across Europe. Cham: Springer; 2021.

Sheremeta L, Knoppers BM. Benefit sharing: it's time for a definition-sharing. In: Philips PWB, Onwuekwa CB, editors. Assessing and sharing the benefits of the genomic revolution. Netherlands: Springer; 2007.

Shickle D, Rhydian H, Carlisle J, et al. Public attitudes to participating in UK biobank: A DNA bank, lifestyle and morbidity database on 500,000 members of the UK public aged 45–69. In: Knoppers BM, editor. Populations and genetics: legal and social-ethical perspectives. Leiden: Martinus Nijhoff Publishers; 2003.

Slokenberga S. Setting the foundations: individual rights, public interest, scientific research and biobanking. GDPR and biobanking. Cham: Springer; 2020.

Ursin LO. Duties and rights of biobank participants: principled autonomy, consent, voluntariness and privacy. In: Solbakk JH, et al., editors. The ethics of research biobanking. New York: Springer; 2009.

Voyer G. Ce que la fréquentation des personnes âgées m'a appris au sujet de l'autonomie ou pour une conception éthique de l'autonomie. In: Barreau du Québec, Autonomie et protection (2007). Cowansville, Que: Éditions Yvon Blais; 2007.

Walker AP. The practice of genetic counseling. In: Uhlmann ES, Schuette JL, Yashar NM, editors. A guide to genetic counseling. 2d ed. Hoboken New Jersey: Wiley-Blackwell; 2009.

Yukl G, Michel JW. Proactive influence tactics and leader member exchange. In: Schriersheim CA, Neider LL, editors. Power and influence in organizations: new empirical and theoretical perspectives. Greenwich: Information Age Publishing; 2001.

Zawati MH. Chapter 12: Liability and the legal duty to inform in research. In: Routledge handbook of medical law and ethics. London: Routledge; 2014.

Secondary Material: Articles

Ahram M, Othman A, Shahrouri M. Public support and consent preference for biomedical research and biobanking in Jordan. Eur J Human Genet 2013;21:567.

Ahram M, Othman A, Shahrouri M, Mustafa E. Factors influencing public participation in biobanking. Eur J Human Genet 2014;22(4):445.

Allen J, McNamara B. Reconsidering the value of consent in biobank research. Bioethics 2011;25(3):155.

Allen C, Joly Y, Moreno PG. Data sharing, biobanks and informed consent: a research paradox? McGill JL Health 2013;7(1):85.

Allen C, Sénécal K, Avard D. Defining the scope of public engagement: examining the 'right not to know' in public health genomics. JL Med Ethics 2014;42:11.

Anand S, et al. Rationale, design, and methods for Canadian alliance for healthy hearts and minds cohort study (CAHHM)—a Pan Canadian cohort study. BMC Public Health 2016;16:650.

Annas GJ. Reforming informed consent to genetic research. J Am Med Assoc 2001;286(18): 2326.

Arias-Diaz J. Spanish regulatory approach for biobanking. Eur J Human Genet 2013;21:708.

Ariss R. The ethic of care in the final report of the Royal Commission on new reproductive technologies. Queen's LJ 1996;22:1.

Arneson RJ. Egalitarianism and the underserving poor. J Political Philos 1997;5:327.

Arribas-Ayllon M. Beyond pessimism: The dialectic of promise and complexity in genomic research. Genom Soc Policy 2010;6(2):1.

Awadalla P, et al. Cohort profile of the CARTaGENE study: Quebec's population-based biobank for public health and personalized genomics. Int'l J Epidemiol 2013;42(5):1285.

Banks TM. Misusing informed consent: a critique of limitations on research subjects' access to genetic research results. Sask L Rev 2000;63:539.

Baron CH. Medical paternalism and the rule of law: a reply to Dr. Relman. Am JL Med 1979; 4:4337.

Barr M. I'm not really read up on genetics: biobanks and the social context of informed consent. BioSocieties 2006;1(2):251.

Bassford HA. The justification of medical paternalism. Social Sci Med 1982;16:731.

Baudoin JL, Parizeau MH. Réflexions juridiques et éthiques sur le consentement au traitement médical. Méd Sci 1987;3:8.

Becker LC. Reciprocity, justice, and disability. Ethics 2005;116(1):9.

Bergmann MM, Moldenhauer J, Borry P. Participants' accounts on their decision to join a cohort study with an attached biobank: a qualitative content analysis study with Two German Studies. J Empirical Res Human Res 2016;11(3):237.

Bélanger-Hardy L. La notion de choix éclairé en droit médical canadien. Health LJ 1997;5:67.

Bélanger-Hardy L. Le consentement aux actes médicaux et le droit à l'autodétermination: développements récents. Ottawa L Rev 1993;25:485.

Berkman BE, Hull SC, Eckstein L. The unintended implications of blurring the line between research and clinical care in a genomic age. Personalized Med 2014;11:285.

Beskow LM, Friedman JY, Hardy NC, Lin L, Weinfurt KP. Developing a simplified consent form for biobanking. PLoS One 2010;5(10):e13302.

Beskow LM, et al. Simplifying informed consent for biorepositories: stakeholder perspectives. Genet Med 2010;12:567.

Beskow LM, et al. Informed consent for population-based research involving genetics. J Am Med Assoc 2001;286:182315.

Beskow LM, Burke W, Fullerton SM, Sharp RR. Offering aggregate results to participants in genomic research: opportunities and challenges. Genet Med 2012;14(4):490−6.

Black V, Gibson E. The temporal element of informed consent. Health L Rev 2004;13:36.

Bollinger JM, Scott J, Dvoskin R, Kaufman D. Public preferences regarding the return of individual genetic research results: findings from a qualitative focus group study. Genet Med 2012;14(4):451.

Borry P, et al. The challenges of the expanded availability of genomic information: an agenda-setting paper. J Commun Genet 2017;1.

Borry P, Shabani M, Howard HC. Is there a right time to know? the right not to know and genetic testing in children. JL Med Ethics. 2014;42:19.

Boniolo G. Trusted consent and research biobanks: towards a 'new alliance' between researchers and doners. Bioethics 2012;26(2):93.

Bowen GA. Document analysis as a qualitative research method. Qual Res J 2009;9(2):27.

Brassington I. John Harris' argument for a duty to research. Bioethics 2007;21:160.

Brassington I. Defending the duty to research? Bioethics 2011;25:21.

Brassington I. The case for a duty to research: not yet proven. J Med Ethics 2014;40:329.

Brazier M. Patient autonomy and consent to treatment: the role of the law? LS 1987;7(2):169.

Bredenoord AL, et al. Disclosure of individual genetic data to research participants: The debate reconsidered. Trends in Genet 2011;27(2):41.

Brekke OA, Sirnes T. Population biobanks: the ethical gravity of informed consent. Bio-Societies 2006;1:385.

Buchanan A. Medical paternalism. Philos Public Affairs 1978;7(4):370.

Buchanan A. Medical paternalism or legal imperialism: not the only alternatives for handling Saikewicz-type cases. Am JL Med 1979;5:97.

Budin-Ljosne I, et al. Dynamic consent: A potential solution to some of the challenges of modern biomedical research. BMC Med Ethics 2017;18(4):1.

Burgess M, O'Doherty KC, Secko D. Biobanking in British Columbia: discussions of the future of personalized medicine through deliberative public engagement. Personalized Med 2008;5(3):285.

Burke W, Evans BJ, Jarvik GP. Return of results: ethical and legal distinctions between research and clinical care. Am J Med Genet 2014;166:105.

Burton PR, et al. Size matters: just how big is big? quantifying realistic sample size requirements for human genome epidemiology. Int'l J Epidemiol 2009;38:263.

Caezanno L, et al. Biobanking research on oncological residual material: a framework between the rights of the individual and the interest of society. BMC Ethics 2013;14:7.

Campbell AV. The ethical challenges of genetic databases: safeguarding altruism and trust. King's LJ 2007;18(22):27.

Campbell A, Glass KC. The legal status of clinical and ethics policies, codes, and guidelines in medical practice and research. McGill LJ 2001;46:473.

Caplan AL. Why autonomy needs help. J Med Ethics 2012:301.

Carnevale FA, et al. Parental involvement in treatment decisions regarding their critically Ill child: a comparative study of France and Quebec. Pediatric Critical Care Med 2007;8(4):337.

Caulfield T, Robertson G. Eugenic policies in Alberta: from the systematic to the systemic? Alta L Rev 1996;59.

Caulfield T. Biobank and blanket consent: the proper place of the public good and public perception rationales. King's LJ. 2007;18(2):209.

Caulfield T, Ries NM. Consent, privacy and confidentiality in longitudinal, population health research: the Canadian legal context. Special Health LJ 2004;12:1.

Caulfield T, Knoppers. Consent, privacy and research biobanks. Policy Brief no. 1 Genomics, Public Policy & Society 1. 2010.

Caulfield T, Murdoch B. Genes, cells, and biobanks: yes, there's still a consent problem. PLoS Biol 2017;15:71.

Caulfield T. Biobanks and blanket consent: the proper place of the public good and public perception rationales. King's LJ. 2007;18(2):209.

Caulfield T, Rachul C, Nelson E. Biobanking, consent, and control: a survey of Albertans on key research ethics issues. Biopreserv Biobank 2012;10(5):433—8.

Chadwick R. The communitarian turn: myth of reality? Camb Q Healthc Ethics 2011;20:546.

Chalmers D, et al. Marking shifts in human research ethics in the development of biobanking. Public Health Ethics 2015;8(1):63.

Chan S, Harris J. Free riders and pious sons — why science research remains obligatory. Bioethics 2009;23:161.

Chandros Hull S, et al. Genetic research involving human biological materials: a need to tailor consent forms. IRB: Ethics Human Res 2004;26(3):1.

Chadwick R, Berg K. Solidarity and equity: new ethical frameworks for genetic databases. Nat Rev Genet 2001;2(4):318.

Charon R. Narrative medicine: a model for empathy, reflection, profession, and trust. J Am Med Assoc 2001;286(15):1897.

Chin JJ. Doctor-patient relationship: from medical paternalism to enhanced autonomy. Singapore Med J 2002;43(3):152.

Christman J. Relational autonomy, liberal individualism, and the social constitution of selves. Philos Stud 2004;117(1/2):143.

Clayton EW. Informed consent and biobanks. JL Med Ethics 2005;33(1):15.

Clayton EW, McGuire AL. The legal risks of returning results of genomics research. Genet Med 2012;14(4):473.

Cohen MB. Perception of power in client/worker relationships. Fam Soc: J Contemp Human Serv 1998;79.

Collins FS. Has the revolution arrived? Nature 2010;464:674.

Collins FS, McKusick VA. Implications of the human genome project for medical science. J Am Med Assoc 2001;285(5):540.

Collins FS, Morgan M, Patrinos A. The human genome project: lessons from large-scale biology. Science 2003;300(5617):286.

Corn BW. Medical paternalism: who knows best? Lancet Oncol 2012;13(2):123.

Coulter A. Paternalism or partnership? British Med J 1999;319:719.

Critchley CR, Nicol D, Otlowski MFA, Stranger MJA. Predicting intention to biobank: a national survey. Eur J Public Health 2010;22:139.

Critchley C, Nicol D, McMhirter R. Identifying public expectations of genetic biobanks. Public Understand Sci 2016;1.

Cross AW, Churchill LR. Ethical and cultural dimensions of informed consent. Ann Internal Med 1982;96(1):110.

D'Ambro F. Biobank research, informed consent and society. towards a new alliance? J Epidemiol Commun Health 2015;1.

D'Ambro F, et al. Research participants' perceptions and views on consent for biobank research: a review of empirical data and ethical analysis. BMC Med Ethics 2016;16:60.

Dauda B, Dierickx K. Benefit sharing: an exploration on the contextual discourse of a changing concept. BMC Med Ethics 2013;14(1):1—8. Avaiable from: http://www.biomedcentral.com/1472-6939/14/36.

Deber RB. Physicians in health care management: 7. the patient-physician partnership: decision making, problem solving and the desire to participate. Can Med Assoc J 1994;151(2):423.

De Clercq B, et al. Health behaviors as a mechanism in the prospective relation between workplace reciprocity and absenteeism: A bridge too far? PLoS One 2015:10—1.

Deschenes M, Sallée C. Accountability in population biobanking: comparative approaches. J Law, Med Ethics 2005;33(1):40.

DeCamp M, Sugarman J. Ethics in population-based genetic research. Account Res 2004;11(1):1.

Del Mar C, Venekamp RP, Sanders S. Antibiotics for children with acute otitis media. J Am Med Assoc 2015;313(151):574.

De Melo-Martin I. A duty to participate in research: does social context matter? AJOB 2008;8:28.

Dhai A, Mahomed S. Biobank research: time for discussion and debate. Issues Med 2013;
103(4):224.

Dickens BM. Medically assisted death: *Nancy B.* v. *Hôtel-Dieu de Québec*. McGill LJ 1993;
38:1053.

Doiron D, Raina P, Fortier I. Linking Canadian population health data: maximizing the potential of cohort and administrative data. Can J Public Health 2013;104(3):e258.

Dove E, et al. Beyond Individualism: Is there a place for relational autonomy in clinical practice and research. Clin Ethics 2017;12(3):150.

Dove ES, Prainsack P. Ethical standards for research biobank donation. J Am Med Assoc 2015;313(15):1574.

Dove ES, Joly Y, Knoppers BM. Power to the people: a wiki-governance model for biobanks. Genom Biol 2012;13(5):1.

Downie J, Llewellyn J. Relational theory & health law and policy. Spec Ed Health LJ 2008;
193.

Downie J, Sherwin S. A feminist exploration of issues around assisted death. Louis U Public L Rev 1996;15(2):303.

Dull MW. Starson v. Swayze, 2003−2008: Appreciating the judicial consequences. Health LJ 2009;17:51.

Dundas I. Case comment: rodriguez and assisted suicide in Canada. Alta L Rev 1994;32:811.

DuVal G. Assisted suicide and the notion of autonomy. Ottawa L Rev 1995;27:1.

Dworkin RB. Getting What we should from the doctors: rethinking patient autonomy and the doctor-patient pelationship. Health Matrix 2003;13:235.

Edwards T. Biobanks containing clinical speicmens: defining characteristics, policies, and practices. Clin Biochem 2014;47:245.

Eiser BJA, Eiser AR, Parmer MA. Power of persuasion: influence tactics for health care leaders. Leadership Action 2006;6:3.

Ells C, Hunt MR, Chambers-Evans J. Relational autonomy as an essential component of patient-centered care. Intl J Feminist Approach Bioet 2011;4(2):79.

Emanuel EJ, Emanuel LL. Four models of the physician-patient relationship. JAMA 1992;
267(16):2221.

Eriksen KÅ, Sundfør B, Karlsson B, Råholm M-B, Arman M. Recognition as a valued human being: perspectives of mental health service users. Nursing Ethics 2012;19(3):357. cited in Sima Sandhu et al. Reciprocity in therapeutic relationships: A conceptual review (2015) 24 International J Mental Health Nursing.

Estey A, Wilkin G, Dossetor J. Are research subjects able to retain the information they are given during the consent process. Health L Rev 1994;3(2):37.

Evans JP, Berg JS. Next-generation DNA sequencing, regulation, and the limits of paternalism. J Am Med Assoc 2011;306(21):2376.

Ewing AT, et al. Demographic differences in willingness to provide broad and narrow consent for biobank research. Biopreserv Biobank 2015;13(2):98.

Facio FM, Brooks S, Loewenstein J, et al. Motivators for participation in a whole-genome sequencing study: implications for translational genomics research. Eur J Human Genet 2011;19(12):1213.

Fenenga CJ, et al. Social capital and active membership in the Ghana national health insurance scheme − a mixed method study. Int'l J Equity Health 2015;14:118.

Fontigny N. Yes really means yes: the law of informed consent in canada revisited. Health L Rev 1996;417.

Forsberg JS, Hansson MG, Eriksson S. Why participating in (certain) scientific research is a moral duty. J Med Ethics 2014;40(5):325.

Forsberg JS, Hansson MG, Eriksson S. Changing perspectives in biobank research: from individual rights to concerns about public health regarding the return of results. Eur J Human Genet 2009;17(12):1544.

Forsberg JS, et al. International guidelines on biobank research leave researchers in ambiguity: why is this so? Eur J Epidemiol 2016;28:449.

Fransson MN, et al. Toward a common language for biobanking. Eur J Human Genet 2015; 23:22.

Gardner AK, Scott DJ. Repaying in kind: Examination of the reciprocity effect in faculty and resident evaluations. J Surgical Edu 2016;1(1):1.

Garrett JR. Ethical considerations for biobaking: Should individual research results be shared with relatives? Futur Med 2012;9(2):159.

Garrison NA, et al. A systematic literature review of individuals' perspectives on broad consent and data sharing in the United States. Genet Med 2015;18(7):663.

Genetics ESoH. Data storage and DNA banking for biomedical research: technical, social and ethical issues. Eur J Human Genet 2003;11(12).

Gessert CE. The problem with autonomy: an overemphasis on patient autonomy results in patients feeling abandoned and physicians feeling frustrated. Minnesota Med 2008; 91(4):40.

Gibson E, et al. Who's minding the shop? the role of Canadian Research Ethics Boards in the creation and uses of registries and biobanks. BMC Med Ethics 2008;9:17.

Gibbons SMC, et al. Governing genetic databases: challenges facing research regulation and practice. JL Soc'y. 2007;34(2):163.

Gillon R. Paternalism and medical ethics. British Med J 1985;290(6486):1971.

Godard B, Marshall J, Laberge C. Community engagement in genetic research: results of the first public consultation for the Quebec CARTaGENE project. Public Health Genom 2007;10(3):147.

Gottweis H, Gaskell G, Starkbaum J. Connecting the public with biobank research: reciprocity matters. Nat Rev Genet 2011;12(11):738.

Gouldner AW. The norm of reciprocity: a preliminary statement. Am Soc Rev 1960:161.

Grady C, et al. Broad consent for research with biological samples: workshop conclusions. Am J Bioeth 2015;15(9):34.

Greely HT. The uneasy ethical and legal underpinnings of large-scale genomic biobanks. Annu Rev Genom Hum Genet 2007;8:343.

Greely HT. Informed consent and other ethical issues in human population genetics. Ann Rev Genom Human Genet 2001;35:785.

Hallinan D, Fiedewald M. Open consent, biobanking and data protection law: can open consent be 'informed' under the forthcoming data protection regulation? Life Sci Soc Policy 2015;11:1.

Halverson CME, Ross LF. Incidental findings of therapeutic misconcemption in biobank-based research. Genet Med 2012;14(6):611.

Hawkins AK. Biobanks: importance, implications and opportunities for genetic counselors. J Genetic Counse 2010;19(5):423.

Hansson MG. Building on relationship of trust in biobank research. J Med Ethics 2005; 31(7):415.

Hansson MG. Ethics and biobanks. British J Cancer 2009;100:8.

Hansson SO. The ethics of biobanks. Camb Q Healthc Ethics 2004;13(4):319.

Harmon SHE. Consent and conflict in medico-legal decision-making at the end of life: a critical issue in the canadian context. UNBLJ 2010;60:208.

Harris JR, Haugan A, Budin-Ljøsne I. Biobanking: from vision to reality. Norsk Epidemiol 2012;21(2).

Harris J. Scientific research is a moral duty. J Med Ethics 2005;31(4):242.

Hartley C. Two conceptions of justice as reciprocity. Social Theory Pract 2014;40(3):409.

Helgesson G. Autonomy, the right not to know, and the right to know personal research results: what rights are there, and who should decide about exceptions? JL Med Ethics 2014;42:28.

Helgesson G. In defense of broad consent. Camb Q Healthc Ethics 2012;21:40.

Hem MH, Pettersen T. Mature care and nursing in psychiatry: Notions regarding reciprocity in asymmetric professional relationships. Health Care Anal 2011;19.

Heins MJ, et al. Effect of the partner's health and support on cancer patients' use of general practitioner care. Psycho-Oncol 2016;25:559.

Hens K, et al. Developing a policy for paediatric biobanks: principles for good practice. Eur J Human Genet 2013;21:2.

Hirschberg I, et al. International requirements for consent in biobank research: qualitative review of research guidelines. J Med Ethics 2014;51:773.

Hobbs A, et al. The privacy-reciprocity connection in biobanking: comparing German with UK strategies. Public Health Genom 2012;15:272.

Hoeyer K. Donor perceptions of consent to and feedback from biobank research: time to acknowledge diversity? Public Health Genom 2010;13(6):345.

Hofmann B. Broadening consent and diluting ethics? J Med Ethics 2009;35:125.

Husak DN. Paternalism and autonomy. Philos Public Affairs 1981:27.

Illingworth P, Parmet WE. The right to health: why it should apply to immigrants. Public Health Ethics 2015;8(2):148.

Ioannidis JPA. Informed consent, big data, and the oxymoron of research that is not research. Am J Bioeth 2013;13(4):40.

Jennings B, Dawson A. Solidarity in the moral imagination of bioethics. Hastings Center Rep 2015;45(5):31.

Johnsson L, et al. Opt-out from biobanks better respects patients' autonomy. British Med J 2008;337:593.

Johnson TC. Reciprocity as a foundation of financial economics. J Business Ethics 2015;131:43.

Jones KH, et al. The other side of the coin: harm due to the non-use of health-related data. Int J Med Informat 2017;97:43.

Joly Y, Dove ES, Knoppers BM, Bobrow M, Chalmers D. Data sharing in the post-genomic world: the experience of the International Cancer Genome Consortium (ICGC) Data Access Compliance Office (DACO). PLoS Comput Biol 2012;8(7):e1002549.

Joly Y, Allen C, Knoppers BM. Open access as benefit sharing? the example of publicly funded large-scale genomic databases. JL Med Ethics. 2012;143.

Joncas D, Philips-Nootens S. Le malentendu thérapeutique: un défi pour le consentement en recherche clinique. RDUS 2005;36:133.

Juth N. The right to know and the duty to tell: the case of relatives. JL Med Ethics 2014;42:38.

Juengst ET, Flatt MA, Settersten RA. Personalized genomic medicine and the rhetoric of empowerment. Hastings Center Rep 2012;42:34.

Jurate S, et al. 'Mirroring' the ethics of biobanking: what analysis of consent documents can tell us? Science Eng Ethics 2014;20:1079.

Katz Jay. Informed consent: a fairy tale? law's vision. U Pitt L Rev 1977;39(2):137.

Kaufman D, et al. Preferences for opt-in and opt-out enrolment and consent models in biobank research: a national survey of veterans' administration patients. Genet Med 2012; 14(9):787.

Kaufman D, Murphy J, Scott J, Hudson K. Subjects matter: a survey of public opinions about a large genetic cohort study. Genet Med 2008;10(11):831.

Kaye J, Whitley EA, Lund D, Morrison M, Teare H, Melham K. Dynamic consent: a patient interface for twenty-first century research networks. Eur J Human Genet 2015;23(2):141.

Kegley JAK. Challenges to informed consent. Eur Mol Biol Org Rep 2004;5(9):832.

Kent A. Consent and confidentiality in genetics: whose information is it anyway? J Med Ethics 2003;29:16.

Kettis-Lindblad A, Lena Ring EV, Hansson MG. Genetic research and donation of tissue samples to biobanks. what do potential sample donors in the swedish general public think? Eur J Public Health 2005;16:433.

Khoury MJ. The case for a global human genome epidemiology initiative. Nat Genet 2004; 36(10):1027.

Kim SYH, et al. Are therapeutic motivation and having one's own doctor as researcher sources of therapeutic misconception? J Med Ethics 2015;41:391.

Kleinman A. The art of medicine — Care: In search of a health agenda. Lancet 2015;386:240.

Knight R, Small W, Shoveller J. How do 'public' values influence individual health behaviour? An empirical-normative analysis of young men's discourse regarding HIV testing practices. Public Health Ethics 2015;1.

Knoppers BM. Population genetics and benefit sharing. Commun Genet 2000;3:212.

Knoppers BM, Chadwick R. Human genetic research: emerging trends in ethics. Nat Rev Genet 2005;6(1):75.

Knoppers BM, Dam A. Return of results: towards a lexicon? JL Med Ethics 2011;39(4):577.

Knoppers BM. Biobanking: international norms. JL Med Ethics 2005;33:7.

Knoppers BM. Consent revisited: points to consider. Health Law Rev 2005;13(2—3):33.

Knoppers BM, Abdul-Rahman (Zawati) MH. Health privacy in genetic research: populations and persons. Politics Life Sci 2009;28:99.

Knoppers BM, Zawati MH, Kirby ES. Sampling populations of humans across the world: ELSI issues. Annu Rev Genom Hum Genet 2012;13:395.

Knoppers BM, Leroux T, Doucet H, Godard B, Laberge C, Stanton-Jean M, et al. Framing genomics, public health research and policy: points to consider. Public Health Genom 2010;13(4):224.

Knoppers BM, Harris JR, Burton PR, Murtagh M, Cox D, Deschênes M, et al. From genomic databases to translation: a call to action. J Med Ethics 2011;37(8):515.

Knoppers BM, Chisholm RL, Kaye J, Cox D, Thorogood A, Burton P, et al. A P3G generic access agreement for population genomic studies. Nature Biotechnol 2013;31(5):384.

Knoppers BM, et al. Towards a data sharing code of conduct for international genomic research. Genom Med 2011;3:46.

Knoppers BM, Dove ES, Litton J-E, Nietfeld J. Questioning the limits of genomic privacy. Am J Hum Genet 2012;91(3):577.

Knoppers BM, Deschênes M, Zawati MnH, Tassé AM. Population studies: return of research results and incidental findings policy statement. Eur J Human Genet 2013;21(3):245.

Knoppers BM. Introduction: from the right to know to the right not to know. JL Med Ethics 2014;42:6.

Knoppers BM, Deschênes M, Zawati MH, Tassé AM. Population studies: return of research results and incidental findings policy statement. Eur J Human Genet 2013;21(3):245.

Kohane IS, et al. Reestablishing the researcher-patient compact. Science 2007;316:836.

Komrad MS. A defense of medical paternalism: maximising patients' autonomy. J Med Ethics 1983;9:38.

Kouri R. L'obligation de renseignement en matière de responsabilité médicale et la «subjectivité rationnelle»: mariage de convenance ou mésalliance? RDUS 1994;24:347.

Kouri R. The law governing human experimentation in Québec. RDUS 1991;22:77.

Kraft SA, Cho MK, Gillespie K, Halley M, Varsava N, Ormond KE, et al. Beyond consent: building trusting relationships with diverse populations in precision medicine research. Am J Bioethics 2018;18(4):3.

Laftman SB, et al. Effort-reward imbalance in the school setting; Associations with somatic pain and self-rated health. Scandinavian J Public Health 2015;43:123.

Laurie G. Evidence of support for biobanking practices. British Med J 2008;337:186.

Laurie G. Recognizing the right not to know: conceptual, professional, and legal implications. JL Med Ethics 2014;42:53.

Laurie G, et al. Managing access to biobanks: how can we reconcile individual privacy and public interests in genetic research? Med Law Int 2010;10:315.

Laurie G. Reflexive governance in biobanking: on the value of policy led approaches and the need to recognise the limits of law. Human Genet 2011;130(3):347.

Lee LM. Adding justice to the clinical and public health ethics arguments for mandatory seasonal influenza immunisation for healthcare workers. Public Health Ethics 2015;41:682.

Lemke AA. Biobank participation and returning research results: perspectives from a deliberative engagement in South Side Chicago. Am J Med Genet 2012;1029.

Lemke AA, Wolf WA, Hebert-Beirne J, Smith ME. Public and biobank participant attitudes toward genetic research participation and data-sharing. Public Health Genom 2010;13:368.

Letendre M, Lanctôt S. Le cadre juridique régissant la relation entre le chercheur et le sujet de recherche: la sécurité conférée par le droit canadien et le droit québécois est-elle illusoire? C de D 2007;48(45):79.

Lévesque A. Chronique: Peut-on consentir à une recherche quand on est un enfant? Psychiatrie, recherche et intervention en santé mentale de l'enfant 1994;4(11):11.

Lévesque E, Knoppers BM, Avard D. La génétique et le cadre juridique applicable au secteur de la santé: examens génétiques, recherche en génétique et soins innovateurs. Revue du Barreau 2004;64:57.

Lévesque E, Joly Y, Simard J. Return of research results: general principles and international perspectives. JL Med Ethics 2011;39(4):583.

Levinson W, et al. Not all patients want to participate in decision making. J General Internal Med 2005;20(6):531.

Lidz CW, Appelbaum PS, Meisel A. Two models of implementing informed consent. Arch Intern Med 1988;148(61):385.

Lidz CW, Appelbaum PS. The therapeutic misconception: problems and solutions. Med Care 2002;40:V55—63.

Lipworth W, Ankely R, Kerridge I. Consent in crisis: the need to reconceptualize consent to tissue banking research. Intern Med J 2006;36:124.

Lipworth W, Forsyth R, Kerridge I. Tissue donation to biobanks: a review of sociological studies. Sociol Health Illness 2011;33(5):792.

Locock L, Boylan AMR. Biosamples as gifts? How participants in biobanking projects talk about donation. Health Expect 2016;19(4):805.

Lunshof JE, et al. From Genetic privacy to open consent. Nat Rev Genet 2008;9:406.

Macneil IR. Exchange Revisited: individual utility and social solidarity. Ethics 1986;96(3):567.

Mahowald MB, Verp MS, Anderson RR. Genetic counselling: clinical and ethical challenges. Ann Rev Genet 1998;32:547.

Marcus-Varwijk AE, et al. Optimizing tailored health promotion for older adults: Understanding their perspectives on healthy living. Gerontol Geriatr Med 2016;2:1.

Marodin G, Braathen Salgueiro J, da Luz Motta M, Pachecho Santos LM. Brazilian guidelines for biorepositories and biobanks of human biological. Mater Rev da Associ Med Brasil 2013;59(1):72.

Marzuk Peter M. The right kind of paternalism. New England J Med 1985;313(23):1474.

Master Z, et al. Biobanks, consent and claims of consensus. Nature Methods 2012;9(9):885.

Matosin N, Frank E, Engel M, Lum JS, Newell KA. Negativity towards negative results: a discussion of the disconnect between scientific worth and scientific culture. Dis Model Mech 2014;7:171.

McCann TV, Clark E. Advancing self-determination with young adults who have schizophrenia. J Psychiatric Mental Health Nursing 2004;11:12.

McCarty CA, et al. Informed consent and subject motivation to participate in a large, population-based genomics study: the marshfield clinic personalized medicine research project. Commun Genet 2007;10:2.

McCoy M. Autonomy, consent, and medical paternalism: legal issues in medical intervention. J Altern Complement Med 2008;14(6):785.

McCullough LB. Was bioethics founded on historical and conceptual mistakes about medical paternalism? Bioethics 2011;25(2):66.

McCullough LB, Wear S. Respect for autonomy and medical paternalism reconsidered. Theor Med Bioethics 1985;6(3):295.

McDonald M. Canadian governance of health research involving human subjects: is anybody minding the store? Health LJ 2001;9:1.

McGregor TL, et al. Inclusion of pediatric samples in an opt-out biorepository linking DNA to De-identified medical records: pediatric BioVU. Clin Pharmacol Ther 2013;93(2):204.

Meier DE, Morrison SR. Autonomy reconsidered. New England J Med 2002;346(14):1087.

Melas PA, et al. Examining the public refusal to consent to DNA biobanking: Empirical data from Swedish population-based study. J Med Ethics 2010;36:93.

Melham K, et al. The evolution of withdrawal: negotiating research relationships in biobanks. Life Sci Soc Policy 2014;10:16.

Melis AP, et al. One for you, one for me: humans' unique turn-taking skills. Psychol Sci 2016;27(7):987.

Merritt M, Grady C. Reciprocity and post-trial access for participants in antiretroviral therapy trials. AIDS 2006;20:1791.

Meslin EM, Cho MK. Research ethics in the era of personalized medicine: updating science's contract with society. Public Health Genom 2010;13(6):378.

Meulenkamp TM, et al. Researchers' opinions towards the communication of results of biobank research: A survey study. Eur J Human Genet 2012;20:258.

Milanovic F, Pontille D, Cambon-Thomsen A. Biobanking and data sharing: a plurality of exchange regimes. Genom Soc Policy 2007;3(1):17.

Molm LD. The structure of reciprocity. Soc Psychol Q 2010;73(2):119.

Molm LD, Schaefer DR, Collet JL. The value of reciprocity. Soc Psychol Q 2007;70(2):199.

Morandeira-Arca J, Etxezarreta-Etxarri E, Azurza-Zubizarreta O, Izagirre-Olaizola J. Social innovation for a new energy model, from theory to action: contributions from the social and solidarity economy in the Basque Country. Innovation: Eur Soc Sci Res 2021:1.

Murphy J, et al. Public perspectives on informed consent for biobanking. Am J Public Health 2009;99(12):2128.

Nelson E, Caulfield T. You can't get there from here: a case comment on *Arndt v. Smith*. UBC L Rev 1998;32:353.

Nelson E, Haymond K, Sidarous M. Selected legal and ethical issues relevant to pediatric genetics. Health LJ 1998;6:83.

Nicol D, Critchley C. Contributing to research via biobanks. Public Understand Sci 2011;1.

Nobile H, Vermeulen E, Thys K, et al. Why do participants enroll in population biobank studies? A systematic literature review. Expert Rev Mol Diagn 2012;13:35.

Novak D, Suzuki E, Kawachi I. Are family, neighbourhood and school social capital associated with higher self-rated health among Croatian high school students? A population-based study. British Med J 2015;5(6):e007184.

Ouellette S, Tassé AM. P3G—10 years of toolbuilding: from the population biobank to the clinic. Appl Transl Genom 2014;3(2):36.

O'Doherty KC, Hawkins AK, Burgess MM. Involving citizens in the ethics of biobank research: informing institutional policy through structured public deliberation. Soc Sci Med 2012;75(9):1605.

O'Neill O. Paternalism and partial autonomy. J Med Ethics 1984;10(4):173.

Ogbogu U, Brown B. Against doctor's orders: the force and limits of personal autonomy in the health care setting. Health LJ 2007;15:515.

Ormond KE, Smith ME, Wolf W. The views of participants in DNA biobanks. Stanford JL Sci Pol'y 2010;1:80.

Osborne PH. Causation and the emerging Canadian doctrine of informed consent to medical treatment. Cases Canadian Law Torts 1985;33:131.

Otlowski MFA. Tackling legal challenges posed by population biobanks: reconceptualising consent requirements. Med L Rev 2012;20:191.

Pieper IJ, Thomson CJ. The value of respect in human research ethics: a conceptual analysis and a practical guide. Monash Bioethics Rev 2014;32(3):232.

Pellegrino ED, Thomasma DC. The conflict between autonomy and beneficence in medical ethics: proposal for a resolution. J Contemp Health L Policy 1987;3:23.

Pers TH, Karjalainen JM, Chan Y, Westra H-J, Wood AR, Yang J, et al. Biological interpretation of genome-wide association studies using predicted gene functions. Nat Commun 2015;6(1):1.

Petersen A. Biobanks' 'engagements': engendering trust or engineering consent? Genom Soc Policy 2007;3(1):31.

Petrini C. 'Broad' consent, exceptions to consent and the question of using biological samples for research purposes different from the initial collection purpose. Social Sci Med 2010; 70:217.

Picard E. Case comment: consent to medical treatment in Canada. Osgoode Hall LJ 1981; 19:140.

Platt J, Bollinger J, Dvoskin R, Kardia SL, Kaufman D. Public preferences regarding informed consent models for participation in population-based genomic research. Genet Med 2014; 16(1):11.

Platt T, et al. 'Cool! and Creepy': Engaging with college student stakeholders in Michigan's biobank. J Commun Genet 2014;5:349.

Pope SJ. What can Christian ethics learn from evolutionary examinations of altruism. J Religion Society 2015;11:138.

Porteri C, Borry P. A proposal for a model of informed consent for the collection, storage and use of biological materials for research purposes. Patient Edu Counse 2008;71(1):136.

Poteri C, Pasqualetti P, Togni E, Parker M. Public's attitudes on particpation in a biobank for research: an Italian survey. BMC Med Ethics 2014;15(1):81.

Prainsack B, Buyx A. A solidarity-based approach to the governance of research biobanks. Med Law Rev 2013;21(1):71.

Prainsack B. The "We" in the "Me": solidarity and health care in the era of personalized medicine. Sci Technol Human Values 2018;43(1):21—44.

Prainsack B, Buyx A. The value of work: addressing the future of work through the lens of solidarity. Bioethics 2018;32(9):585—92.

Pray L. Personalized medicine: hope or hype? Nat Edu 2008;1:72.

Pullman D. Subject comprehension, standards of information disclosure and potential liability in research. Health LJ 2001;9:113.

Pullman D, Etchegary H, Gallagher K, Hodgkinson K, Keough M, Morgan D, Street C. Personal privacy, public benefits, and biobanks: a conjoint analysis of policy priorities and public perceptions. Genet Med 2012;14:229.

Qian J, et al. Mental health risks among nurses under abusive supervision: the moderating roles of job role ambiguity and patients' lack of reciprocity. Int'l J Mental Health Syst 2015;9(22):1.

Quill TE, Brody H. Physician recommendations and patient autonomy: finding a balance between physician power and patient choice. Ann Intern Med 1996;125:763.

Rahm AK, Wrenn M, Carroll NM, Feigelson HS. Biobanking for research: a survey of patient population attitudes and understanding. J Commun Genet 2013;4(44):45.

Rantanen E, et al. What is ideal genetic counselling? a survey of current international guidelines. Eur J Human Genet 2008;16(4):445.

Ravitsky V, Wilfond B. Disclosing individual genetic results to research participants. AJOB 2006;6(6):8.

Rich BA. Medical paternalism v. respect for patient autonomy: the more things change the more they remain the same. Michigan State University J Med L 2006;10:87.

Richter G, et al. Broad consent for health care-embedded biobanking: understanding and reasons to donate in a large patient sample. Genet Med 2018;20(1):76.

Roberston GB. Informed consent in Canada: an empirical study. Osgoode Hall LJ 1984; 22:139.

Robertson GB/ Ontario's new informed consent law: codification or radical change? Health LJ 1994;2:88.

Robertson GB. Informed consent ten years later: the impact of *Reibl v. Hughes*. Can Bar Rev 1991;70:423.

Rodgers-Magnet S. Recent developments in the doctrine of informed consent to medical treatment Re: *Hopp v. Lepp* and *Reibl v. Hughes*. Cases Canadian Law Torts 1980;14:61.

Rodriguez H, Snyder M, Uhlén M, Andrews P, Beavis R, Borchers C, et al. Recommendations from the 2008 international summit on proteomics data release and sharing policy: the Amsterdam principles. J Proteom Res 2009;8(7):3689.

Rodriguez-Osorio CA, Dominguez-Cherit G. Medial decision making: paternalism versus patient-centred (autonomous) care. Curr Opin Crit Care 2008;14:708.

Rhodes R. Rethinking research ethics. AJOB 2005;7:7.

Ross L. Phase I research and the meaning of direct benefit. J Pediatr Suppl 2006;149:S20.

Saas HM. Advance directives for psychiatric patients? balancing paternalism and autonomy. WMV Wiener Medizinische Wochenschrift 2003;153(17):380.

Sak J, Pawlikowski J, Goniewicz M, Witt M. Population biobanking in selected European countries and proposed model for a polish national DNA bank. J Appl Genet 2012;53:159.

Sanderson SC. Genome sequencing for healthy individuals. Trends Genet 2013;29(10):556.

Sanderson SC, et al. Willingness to participate in genomics research and desire for personal results among underrepresented minority patients: a structured interview study. J Commun Genet 2013;4(4):469.

Sandhu S, et al. Reciprocity in therapeutic relationships: a conceptual review. Int'l J Mental Health Nursing 2015;24:460.

Sandman L, Munthe C. Shared decision-making and patient autonomy. Theoretical Med Bioethics 2009;30(4):289.

Sarojini S, et al. Proactive biobanking to improve research and health care. J Tissue Sci Eng 2012;3(2):116.

Schroeder D. Benefit sharing: it's time for a definition. J Med Ethics 2007;33:205.

Secker B. The appearance of kant's deontology in contemporary kantianism: concepts of patient autonomy in bioethics. J Med Philos 1999;24(1):43.

Siegler M. The progression of medicine: from physician paternalism to patient autonomy to bureaucratic parsimony. Arch Intern Med 1985;145:713.

Shabani M, Knoppers BM, Borry P. From the principles of genomic data sharing to the practices of data access committees. EMBO Molecu Med 2015;7(5):507.

Shabani M, Borry P. You want the right amount of oversight: interviews with data access committee members and experts on genomic data access. Genet Med 2016;18(9):892.

Shabani M, Dyke SO, Joly Y, Borry P. Controlled access under review: improving the governance of genomic data access. PLoS Biol 2015;13(12):e1002339.

Shabani M, Dove ES, Murtagh M, Knoppers BM, Borry P. Oversight of genomic data sharing: what roles for ethics and data access committees? Biopreserv Biobank 2017;15(5):469.

Shapshay S, Pimple KD. Participation in biomedical research is an imperfect moral duty: a response to John Harris. J Med Ethics 2007;33:414.

Sharpe N. Reinventing the wheel?: informed consent and genetic testing for breast cancer, cystic fibrosis, and huntington disease. Queen's LJ 1997;22:389.

Shaw DM, Elger BS, Colledge F. What is a biobank? differing definitions among biobank stakeholders. Clin Genet 2014;85(3):223.

Shickle D. The consent problem within DNA biobanks. Stud Hist Philos Sci Biol Biomed Sci 2006;37(3):503.

Siegler M. The progression of medicine: from physician paternalism to patient autonomy to bureaucratic parsimony. Arch Intern Med 1985;145(4):713.

Silva DS, Dawson A, Upshur REG. Reciprocity and ethical tuberculosis treatment and control. Bioethical Inquiry 2016;13:75.

Simm K. Benefit-sharing: an inquiry regarding the meaning and limits of the concept in human genetic research. Genom Soc Policy 2005;1:29.

Skipper M. The peopling of Britain. Nat Rev Genet 2015;16:256.

Smith MJ. Population-based genetic studies: informed consent and confidentiality. Santa Clara Computer High Tech LJ 2001;18:57.

Solberg B, Steinsbekk KS. Biobank consent models — are we moving toward increased participant engagement in biobanking? J Bioreposit Sci Appl Med 2015;3:23.

Solberg B, Steinsbekk KS. Managing incidental findings in population based biobank research. Norsk Epidemiologi 2012;21(2):195.

Stauton C, Moodley K. Callenges in biobank governance in Sub-Saharan Africa. BMC Med Ethics 2013;14:35.

Stein DT, Terry SF. Reforming biobank consent policy: a necessary move away from broad consent toward dynamic consent. Biopreserv Biobank 2013;17(12):855.

Steinsbekk KS, Myskja BK, Solberg B. Broad consent *versus* dynamic consent in biobank research: is passive participation an ethical problem? Eur J Human Genet 2013;21(9):897.

Stirrat GM, Gill R. Autonomy in medical ethics after o'neill. J Med Ethics 2005;31:127.

Stretch D. Ethical review of biobank research: should recs review each release of material from biobanks operating under an already-approved broad consent and data protection model? Eur J Med Genet 2015;58:545.

Stretch D, et al. A template for broad consent in biobank research. results and explanation of an evidence and consensus-based development process. Eur J Med Genet 2016;59:295.

Sudlow C, Gallacher J, Allen N, Beral V, Burton P, Danesh J, et al. UK biobank: an open access resource for identifying the causes of a wide range of complex diseases of middle and old age. PLos Med 2015;12(3):e1001779.

Sutrop M. How to avoid a dichotomy between autonomy and beneficence: from liberalism to communitarianism and beyond. J Internal Med 2011;269(4):275.

Stephens C, Breheny M, Mansvelt J. Volunteering as reciprocity: beneficial and harmful effect of social policies to encourage contribution in older age. J Aging Studies 2015;33:22.

Sutrop M. Changing ethical frameworks: from individual rights to the common good? Camb Q Healthc Ethics 2011;20:533.

Swede H, Stone CL, Norwood AR. National population-based biobanks for genetic research. Genet Med 2007;9(3):141.

Tan N. Deconstructing paternalism: what serves the patient best. Singapore Med J 2002; 43(3):148.

Tassé AM, Budin-Ljøsne I, Knoppers BM, Harris JR. Retrospective access to data: the ENGAGE consent experience. Eur J Human Genet 2010;18(7):741.

Taylor K. Paternalism, participation and partnership—the evolution of patient centeredness in the consultation. Patient Edu Counse 2009;74(2):150.

Tazzioli M, Walters W. Migration, solidarity and the limits of Europe. Global Discourse 2019; 9(1):175–90.

Teare HJA, Morrison M, Whitley EA, Kaye J. Towards 'Engagement 2.0': Insights form a study of dynamic consent with biobank participants. Digital Health 2015;0(0):1.

Theil DB, et al. Testing an online, dynamic consent portal for large population biobank research. Public Health Genom 2015;18:26.

Thomasma DC. Beyond medical paternalism and patient autonomy: a model of physician conscience for the physician-patient relationship. Ann Intern Med 1983;98:243.

Thornton H. The UK biobank project: trust and altruism are alive and well: a model for achieving public support for research using personal data. Int J Surgery 2009;7:501.

Thorogood A, Joly Y, Knoppers BM, Nilsson T, Metrakos P, Lazaris A, et al. An implementation framework for the feedback of individual research results and incidental findings in research. BMC Med Ethics 2014;15(1):1.

Tinetti ME, Basch E. Patients' responsibility to participate in decision making and research. JAMA 2013;309:2331.

Tomlinson T. Respecting donors to biobank research. Hastings Center Rep 2013;43(1):41.

Tomlinson T, et al. Moral concerns and the willingness to donate to a research biobank. JAMA 2015;313(4):417.

Toronto International Data Release Workshop Authors. Prepublication data sharing. Nature 2009;461:168.

Torrens J. Informed consent and the learned intermediary rule in Canada. Sask L Rev 1994;58: 399.

Tôth F. Le droit du patient d'être informé: un droit protégé par la Charte des droits et libertés de la personne. RDUS 1989;20:161.

Tu JV, et al. Impracticability of informed consent in the registry of the Canadian Stroke Network. New England J Med 2004;350:1414.

Tupasela A. From gift to waste: changing policies in biobanking practices. Sci Public Policy 2011;38(7):510.

Ursin LO. Personal autonomy and informed consent. Med Health Care Philos 2009;12(1):17.

Veatch RM. Models for ethical medicine in a revolutionary age. The Hastings Center Rep 1972;2(3):5.

Venkat A, et al. Ethical issues in the response to Ebola virus disease in United States emergency departments: a position paper of the American College of Emergency Physician, the Emergency Nurses Association, and the Society for Academic Emergency Medicine. Acad Emerg Med 2015;22(5):605.

Viens AM. Public health, ethical behavior and reciprocity. Am J Bioethics 2008;8(5):1.

Virani AH, Longstaff H. Ethical considerations in biobanks: how a public health ethics perspective sheds new light on old controversies. J Genetic Counselling 2015;24:428.

Walker RL. Medical ethics needs a new view of autonomy. J Med Philos 2009;33(6):594.

Walport M, Brest P. Sharing research data to improve public health. Lancet 2011;377:537.

Walter JK. Relational autonomy: moving beyond the limits of isolated individualism. Pediatrics 2014;133:S16.

Weiss GB. Paternalism modernised. J Med Ethics 1985;11:184.

Weisstub DN. Roles and fictions in clinical and research ethics. Health LJ 1996;4:259.

Weisstub DN, Verdun-Jones SN. Pour une distinction entre l'expérimentation thérapeutique et l'expérimentation non thérapeutique. RDUS 1997;27:49.

Wendler D. Broad versus blanket consent for research with human biological samples. Hastings Centre Rep 2013;43(5):3.

Wendler D. One-time general consent for research biological samples. British Med J 2006; 332:544.

Wertz DC. Patient and professional views on autonomy: a survey in the United States and Canada. Health Law Rev 1998;7(3):9.

Williams G, Schroeder D. Human genetic banking: altruism, benefit and consent. New Genet Soc 2004;23:89.

Wolf SM, Paradise J, Caga-anan C. The law of incidental findings in human subjects research: establishing researchers' duties. JL Med Ethics 2008;36(2):361.

Wolf SM, et al. Managing incidental findings and research results in genomic research involving biobanks and archived data sets. Genet Med 2012;14:361.

Woods S, McCormack P. Disputing the ethics of research: the challenge from bioethics and patient activism to the interpretation of the declaration of helsinki in clinical trials. Bioethics 2012.

Zaidi MY, Haddad L, Lathrop E. Global health opportunities in obstetrics and gynecology training: Examining engagement through an ethical lens. Am J Tropical Med Hygiene 2015;93(6):1194.

Zawati MH. There will be sharing: population biobanks, the duty to inform and the limitations of the individualistic conception of autonomy. Health LJ 2014;21:97.

Zawati MnH, Rioux A. Biobanks and the return of research results: out with the old and in with the new? JL Med Ethics 2011;39(4):614.

Zawati MnH, Knoppers BM. International normative perspectives on the return of individual research results and incidental findings in genomic biobanks. Genet Med 2012;14(4):484.

Zimmern J. Consent and autonomy in the human tissue act 2004. King's LJ 2007;18:313.

Other Materials

Atlantic PATH. Our study. 2018. http://atlanticpath.ca/.

Atlantic PATH. Consent and brochure (obtained through correspondence). 2013.

BC Generations Project. The project. 2019. http://www.bcgenerationsproject.ca/.

BC Generations Project. Consent Form. British Columbia; 2014 (obtained through correspondence).

Biobanking and biomolecular resources research infrastructure (BBMRI) BBMRI–ERIC. 2018. http://www.bbmri-eric.eu/.

Canadian alliance for healthy hearts & minds. Timetable: Release of Data; 2017. http://cahhm.mcmaster.ca/?page_id=4278.

Canadian Alliance for Healthy Hearts and Minds. Participant information and consent sheet (aboriginal participants), (obtained through correspondence).

Canadian alliance for healthy hearts and minds, participant information and consent sheet (MHI site) (obtained through correspondence).

Canadian Institutes of Health Research. Canadian longitudinal study on aging (CLSA). Government of Canada; 2018. http://www.cihr-irsc.gc.ca/e/18542.html.

Canadian Longitudinal Study on Aging. About the study. 2018. online: www.clsa-elcv.ca/.

Canadian Longitudinal Study on Aging. Governance. 2018. www.clsa-elcv.ca/about-us/governance.

Canadian Longitudinal Study on Aging. Study information package – Home interview & data collection site visit. www.clsa-elcv.ca/doc/414.

Cancer Care Manitoba. CCMB tomorrow project. 2018. http://www.cancercare.mb.ca/resource/File/CCMB-Tmrw-Proj_pamphlet_FNL_R1_web.pdf.

CanPath Portal. Canadian partnership for tomorrow project. 2018. www.partnershipfortomorrow.ca.

CanPath Portal. Alberta's tomorrow project (Alberta). 2015 (recruitment statistic updated as of February 2015), https://portal.canpath.ca/mica/individual-study/atp.

CanPath's Access Portal Documents. Data access policy, a publications policy, an intellectual property policy and a data access application form, https://portal.canpath.ca/user/login?destination=node/7.

CanPath Portal. Leadership and governance. 2020. https://canpath.notadev.site/governance/.

CanPath Portal. Data and samples access application form. 2016. https://portal.partnershipfortomorrow.ca/agate/register/#/join.

CanPath Portal. Publications policy, https://portal.canpath.ca/sites/live-7x35-1-release-15976 81236-mr-portal/files/CanPath%20Publications%20Policy%20PDF%20Version.pdf.

CanPath Portal. Approved projects. 2015. https://portal.partnershipfortomorrow.ca/mica/research/projects.

Canadian Tumour Repository Network (CTRNet) Website. Standard operating procedures, www.ctrnet.ca/operating-procedures.

CARTaGENE. About. 2016. https://cartagene.qc.ca/en/about.

CARTaGENE. Second wave information brochure for participants. 2014. https://cartagene.qc.ca/sites/default/files/documents/consent/cag_2e_vague_brochure_en_v3_7apr2014.pdf.

CARTaGENE Twitter Account. https://twitter.com/_cartagene_?lang=en.

Council of Canadian Academies. Accessing health and health-related data in Canada: executive summary. Ottawa: Council of Canadian Academies; 2015. http://www.scienceadvice.ca/uploads/eng/assessments%20and%20publications%20and%20news%20releases/Health-data/HealthDataExecSumEn.pdf.

Count me in 4 Tomorrow. Brief history & summary. 2012. http://in4tomorrow.ca/.

Dworkin G. Paternalism. In: Zalta EN, editor. Stanford encyclopedia of philosophy. Winter; 2017. Available from: plato.stanford.edu/entries/paternalism/.

Estonian Genome Centre. University of Tartu, <www.geenivaramu.ee/en/>.

EMBL-EBI. What is proteomics? 2021. https://www.ebi.ac.uk/training/online/courses/proteomics-an-introduction/what-is-proteomics/.

EurekAlert!. Kaiser Permanente study: Alcohol amount, not type—wine, beer, liquor—triggers breast cancer. www.eurekalert.org/pub_releases/2007-09/kpdo-kps092207.php.

Fonds de la recherche en santé du Québec. Final Report — Advisory Group on a governance framework for data banks and biobanks used for health research. 2006. www.frsq.gouv.qc.ca/en/ethique/pdfs_ethique/Rapport_groupe_conseil_anglais.pdf.

Guion LA, Diehl DC, McDonald D. Triangulation: establishing the validity of qualitative studies. EDIS 2011;2011(8):3.

Healthy Ageing Campus Groningen. Healthy Ageing Campus. 2018. https://campus.groningen.nl/about-campus-groningen/healthy-ageing-campus.

HumGen International. HumGen Database: your resource in ethical, legal and social issues in human genetics. 2018. www.humgen.org.

McGill University — Faculty of Medicine. General guidelines for biobanks and associated databases. 2015. www.mcgill.ca/medresearch/files/medresearch/guidelines_for_biobanks_and_associated_databases.march2015.pdf.

New York Times. For women, confusion about alcohol and health. Parker-Pope T; 9 Oct. 2007. https://well.blogs.nytimes.com/2007/10/09/at-cocktail-time-shots-of-confusion/.

Nuremberg Military Tribunals. Permissible medical experiments. In: Trials of war criminals before the Nuremberg military tribunals under control council law, vol 10:2. Washington: US Government Printing Office; 1949.

Office for Human Research Protections. U.S. Department of Health and Human Services, International Compilation of Human Research Standards. Washington: Department of Health and Human Services; 2020.

Ontario Health Study. About the study. 2018. www.ontariohealthstudy.ca/.

Ontario Health Study. Website FAQ. 2014. www.ontariohealthstudy.ca/about-the-study/frequently-asked-questions/.

Ontario Health Study. Consent Form (obtained through correspondence). 2014.

Oxford English Dictionary. Online Edition, www.oed.com.

PopGen International Database. Population Biobanks Lexicon, a collaborative endeavour between: Public Population Project in Genomics and Society (P3G) & Promoting Harmonization of Epidemiological Biobanks in Europe (PHOEBE), Glossary: biobank, http://www.popgen.info/glossary.

The International Cancer Genome Consortium's Data Access Application Office–DACO. DACO approved projects. 2021. https://daco.icgc.org/controlled-data-users/#/?&order= asc&by=proj.

The Tomorrow Project. Consent Form. 2011. Alberta (obtained through correspondence).

The Tomorrow Project. Study Booklet. 2011. Alberta (obtained through correspondence).

UK Biobank. UK Biobank. 2018. www.ukbiobank.ac.uk.

UK Biobank. Newsletter. 2015. 2015 [archived website], https://web.archive.org/web/ 20200929155649/http://www.ukbiobank.ac.uk/newsletter-2015/.

US National Library of Medicine. Learn about clinical studies. 2017. www.clinicaltrials.gov/ ct2/about-studies/learn#ClinicalTrials.

Wellcome Trust. Enabling data linkage to maximise the value of public health researh data: full report. 2015. https://cms.wellcome.org/sites/default/files/enabling-data-linkage-to-maximise-value-of-public-health-research-data-phrdf-mar15.pdf.

Wellcome Trust. Signatories to the joint statement, https://wellcome.org/what-we-do/our-work/sharing-research-data-improve-public-health-full-joint-statement-funders-health#the-joint-statement-of-purpose-5ea3.

Index

Printed in the United States
by Baker & Taylor Publisher Services